Between scientism and religious fundamentalism lie many more-thoughtful options. Four of them, developed from the perspective of four disciplines, are attractively offered in this book. The book makes it abundantly clear that there is no reason for opposition between science and faith. It successfully stimulates serious thought about the real, and complex, issue of how best to relate them.

—John B. Cobb Jr., professor emeritus,
Claremont School of Theology
Author of *Back to Darwin: A Richer Account of Evolution*

The issues surrounding science and theology are multifaceted. Most books dealing with these issues provide one point of view. This book does it right! Four authors deal separately with the biblical accounts of creation, evolutionary theory, philosophical influences on evolutionary thought, and finally, a positive transcendence of the false dichotomies with an exciting conclusion: "... reality is a grand journey, and God is the ultimate adventurer." It is a brilliant resolution to a difficult issue.

—Roger S. Fouts, PhD, Dean, Graduate Studies and Research
Professor of psychology, Central Washington University
Author of *Next of Kin: My Conversations with Chimpanzees*

AUTHOR ACKNOWLEDGMENTS

We wish to thank our editor, Jerry Ruff, whose professionalism and insight have made this project possible.

PUBLISHER ACKNOWLEDGMENTS

Thank-you to the following individuals who reviewed this work in progress:

Eugene Selk
Creighton University, Omaha, Nebraska

Gary Stansell
St. Olaf College, Northfield, Minnesota

Shannon Schrein
Lourdes College, Sylvania, Ohio

GENESIS EVOLUTION AND THE SEARCH FOR A REASONED FAITH

Mary Katherine Birge, SSJ
Brian G. Henning
Rodica M. M. Stoicoiu
Ryan Taylor

Created by the publishing team of Anselm Academic.

Cover art royalty free from iStock

Printed in the United States of America

7031 (PO2844)

ISBN 978-0-88489-755-2

To our students

CONTENTS

INTRODUCTION

In 2009 the world rightly paused amid the calamitous cacophony of the global economic downturn and ongoing environmental crisis to celebrate the 200th anniversary of the birth of Charles Darwin (1809–1882). The year also marked the 150th anniversary of the publishing of Darwin's revolutionary *On the Origin of Species by Means of Natural Selection* (1859). This dual cause for celebration brought many to reflect on the contributions of evolutionary theory to humanity's understanding of itself, its place in the cosmos, and its relationship to the transcendent.

As is characteristic of this day and age, the public discussion of Darwin's great insight has not always enjoyed a nuanced treatment in the mainstream media. Given the popular presentation of evolution as a sort of celebrity death match between religion and science (*Only one can leave the ring alive!*), it is no wonder that many people think they must choose between religion and science, faith and reason, Genesis and evolution. Indeed, this is just what the contributors to this book have found in their classrooms; too frequently students appear to live in an intellectually and spiritually bifurcated world in which they must pick either evolution or creation and shun the other or hold both without considering how they work together. It is out of this fraught context that this book was born.

The idea for this project grew out of ongoing forums on religion and evolution coordinated by Rodica Stoicoiu at Mount St. Mary's University in Emmitsburg, Maryland.[1] In these informal roundtables, students come together with faculty from science, philosophy, and theology to discuss and debate the intersection of seemingly conflicting ideas around evolutionary biology and the Christian faith. Among the faculty participants have been the authors of this text, Mary Katherine Birge, SSJ (biblical studies), Brian G. Henning (philosophy), Rodica Stoicoiu (systematic theology), and Ryan Taylor (evolutionary biology). Despite their diverse disciplinary perspectives

1. At that time (2006), all of the authors were teaching at Mount St. Mary's University. Taylor is now teaching at Salisbury University in Salisbury, Maryland, and Henning is now at Gonzaga University in Spokane, Washington.

and training, the authors each realized that they reject a false dichotomy between faith and science. The guiding principle of this text is that a thoughtful individual need not choose between the two; there is a way to proceed through the quagmire of well-intended presumptions about science and faith, and, more specifically, the theory of evolution and the creation stories in the Book of Genesis. That way is a dialogue among disciplines. Rather than eschew nuance and gloss over complexity, *Genesis, Evolution, and the Search for a Reasoned Faith* is the authors' attempt to bring together truths revealed by evolutionary biology and religious faith. In an important sense, this volume is the authors' joint attempt to model the sort of discussion their students deserve to hear.

The simple structure of the text is intended to mirror this dialogical impetus. In chapter 1, biblical scholar Birge examines Genesis 1–3, exploring when it was written, who wrote it, what was going on in the world of the authors and their audiences at the time it was written, what those authors may have intended their work to mean to those ancient audiences, and how a modern audience may understand with a reasoned faith what the texts have to say. Birge notes that *reading* what Genesis *says* is not the same as *understanding* what the text *means*, though biblical literalists would suggest otherwise. Approaching the creation accounts within the context of the rich and complex history of the Isrealite people reveals that Genesis is not a scientific treatise giving a play-by-play account of how God created the universe. Rather, the creation stories in Genesis are a deeply theological exploration of how human beings should see their relationship to a transcendent creator. Taken in this vein, one may see evolutionary science and religious faith as complementary, not contradictory, attempts to understand humanity's origins.

In chapter 2, biologist Taylor begins by exploring the often-misunderstood nature of scientific investigation, focusing in particular on what scientists mean when they talk about a scientific "fact" or a scientific "theory." While in everyday usage *theory* might mean little more than a formulated opinion or guess, in science *theory* denotes a hypothesis (tentative explanation) that has never failed to be confirmed by empirical testing and observation—hardly a mere opinion.

Recognizing the empirical, inductive basis of all scientific investigation, Taylor notes that science cannot ask, much less answer, questions concerning the meaning of human existence, or whether there is a supernatural creator. Take, for example, a hypothesis that longer-legged deer in a particular deer population have a "leg up" on their companions in the struggle for survival. This is a question that is open to scientific study. "Does God exist?" on the other hand, is not such a question. Scientific hypotheses must be testable questions that can either be supported or proved wrong. While scientists can design a series of experiments to test the deer hypothesis, the question of God's existence does not lend itself to such experimentation. As the body of data testing a scientific question builds over time and confidence climbs to ever higher degrees of certainty (though science never claims to be completely certain), these hypotheses come to be considered theories, as close to certainty as science can get. Evolution is one such theory.

The modern synthesis of the theory of evolution by natural selection, which takes into account the role of genetics, is accepted by most scientists as the unifying conceptual framework that explains the origins of our species, *Homo sapiens*, and the millions of other life-forms on our planet. Yet, Taylor notes, the methodological naturalism of evolutionary theory requires that scientists remain silent regarding transcendent questions. Questions concerning the meaning of human life or the existence of a transcendent creator must be left to philosophers and theologians.

Picking up these questions in chapter 3, philosopher Henning explores the ethical and philosophical significance of the theory of evolution by tracing the history of ideas that led up to and beyond Darwin's great discovery. This philosophical investigation leads Henning to ask such questions as, "Does modern evolutionary theory adequately explain the origins of consciousness?" "Is it possible for conscious beings to evolve from completely lifeless and mindless matter?" "Does the recognition of humanity's shared evolutionary heritage undermine our human-centered worldview, or require that we change, particularly with respect to how we treat nonhuman life?"

Henning notes the strong tendency in Western thought to place humans at the top of a hierarchy of being. Modern evolutionary theory fundamentally challenges the assumption that humans are utterly unique. Rather than being at the pinnacle of creation, distinct from all other life-forms, the theory of evolution places humans on a continuum of being, a continuum that challenges the idea that those things that make us who we are, such as culture, language, reason, and so on, are unique to us. The theory of evolution opens the door to the idea that those beings from whom we developed and those that are genetically close to us today may hold these same characteristics, though perhaps to different degrees. Rather than being a singular exception to the forces that shaped the natural world, human beings are a great exemplification of such forces.

In recognizing this, Henning notes that evolutionary biology in turn must abandon the notion that physical reality is best understood as a valueless machine, deterministically playing out its programming. If, as evolutionary science teaches, humans evolved from simpler organisms, and if human beings are subjects who are free, conscious, and (at least intermittently) self-reflective, then this sense of freedom and subjectivity also must be found in humanity's evolutionary ancestors.

In the fourth and last chapter, systematic theologian Stoicoiu seeks to interweave the threads of conversation from the preceding chapters and demonstrate the fundamental intellectual inadequacy of not only atheistic evolutionary materialism and simplistic biblical creationism but also more sophisticated contemporary approaches, such as scientific creationism and intelligent design theory. Rather than seeing the theory of evolution as a threat to religious belief, Stoicoiu suggests that a theology that embraces evolution can deepen and broaden a faith seeking understanding. In this way, she rejects the impulse to save religion by retreating into "separatism" (the view that science and religion are nonoverlapping domains of inquiry). From the perspective of biblical creation stories, one can come to understand how these stories answer important transcendental questions, while realizing that one cannot also expect them to address questions posed by modern science. Today, one can build upon biblical creation accounts and, with the help of theology, address evolutionary theory, not as some construct that lies outside the theological sphere, but rather as a theory to be theologically engaged.

Stoicoiu concludes that one must respect the autonomy and veracity of evolutionary biology, recognize the reality and ubiquity of suffering in the world, and begin to move toward an evolutionary theology that recognizes the richness that evolutionary theory can bring to one's understanding of the transcendent's relationship to creation. One of the great lessons theology can glean from a study of evolution is that all of reality is in the process of becoming. Theology recognizes this process and sees in it the means of drawing closer to the mystery of God. In this light, evolution is constantly offering us a world in transformation. Theology understands this transformation in light of a hope-filled promise of the future when the fulfillment of God's word will be realized. In the end, we need not choose between religion or science, faith or reason, Genesis or evolution. Evolution is not a threat to faith, but rather an enrichment of faith. A thorough faith seeking understanding brings together Genesis *and* evolution.

Mary Katherine Birge, SSJ
Brian G. Henning
Rodica M. M. Stoicoiu
Ryan Taylor

chapter 1

Genesis

Mary Katherine Birge, SSJ
Mount St. Mary's University, Emmitsburg, MD

Before beginning to read any part of the Bible, one ought to ask several important questions. What is the Bible? Who wrote it? Where and when was it written? For whom was it written, and why? These questions and their answers will generate further questions, such as: What was happening in the world of the writer and the writer's audience when the Bible was being written? What literary form did the author choose to write in and why? What do people mean when they say the Bible is "inspired"? Having answered such queries, one may begin to explore individual books of the Bible, including Genesis and its creation accounts. What do these accounts say to a modern audience about God, humanity, and science, and the truth claims that various, and sometimes competing, constituencies have made for them?

WHY READ THE BIBLE IN THE FIRST PLACE?

The Bible is not simply one book, as indicated by the Greek origin of its name, *ta biblia*, "the books." These books have been divided into two distinct sections, the Old and New Testaments. Today, people read and study the Bible for many reasons. The Bible helped shape the last two millennia of Western society in terms of history, government, literature, art, music, architecture, philosophy, and religion. For that reason alone, anyone who lives or works in the West needs some knowledge and understanding of major biblical stories and images to be conversant with a cultural force that has shaped, and continues to shape, modern worldviews and activities. For many others, however, the Bible is much more than a significant cultural artifact; for Christians and Jews alike, the Bible contains the word of God.[1]

Throughout the last three millennia, Jews and Christians have preserved biblical texts because in them they find wisdom, understanding, personal guidance, and enlightenment regarding their own experiences of God. Their underlying assumption about the Bible is that God wants to communicate with them.

For Jews and Christians, one of the privileged places where God continues to reveal the Divine's hopes and dreams for humanity is the Bible. How such believers proceed to read and interpret these sacred texts will rest on their understanding of how God collaborated with the human creators of individual stories, collections of stories, and whole books of the Bible. For most modern Jews and Christians, God continues to imbue their sacred stories with wisdom and guidance directly related to the wisdom and guidance the ancient authors and their audiences received. For these believers, just as God participated with the human community in the initial collecting, preserving, and sharing of biblical texts, so God continues to participate as they interpret the text and apply its message with each new generation.

Biblical Inspiration: What's God Got to Do with It?

In 2 Timothy 3:16 we read, "All scripture is inspired by God." With this verse, the author expresses his belief that the storytellers, composers, writers, compilers, and editors of biblical texts were all assisted by God. The verb *to inspire* has its roots in the Latin verb *inspirare*,

1. When Christians use the word *Bible*, ordinarily they mean both the Old and the New Testaments. Jews use the terms *Bible* or *Tanak* [an acronym from the Hebrew words for Law (*Torah*), Prophets (*Nebi'im*), and Writings (*Ketubim*)], for the collection of books that Christians call the Old Testament. For the purposes of this chapter, the word *Bible* indicates the collection of books found simultaneously in the *Tanak* and in the Old Testament.

which means "to breathe into." In some way, then, the author of 2 Timothy believes that God has "breathed" God's own self "into" every stage of the Bible's composition, with the effect that in all **Scripture** is "such truth as God, for the sake of our salvation, wished the biblical text to contain."[2] What is true, that is, what is inerrant or without error in the Bible, are those things that are necessary for the believer's salvation. Biblical texts also may include information non-essential to salvation, information that contemporary readers have rightly determined to be erroneous in matters of geography, history, science, or mathematics.

CHRISTIAN FUNDAMENTALISM

Some Christians reject any limits on the Bible's **inerrancy**. They insist that the truth of Scripture includes every word in its literal sense, a notion called **plenary verbal inerrancy** or **strict inerrancy**. These Christians form a movement called **fundamentalism**, which began in the United States in the late nineteenth and early twentieth centuries in response to a perceived threat to the "fundamental" truths of the Bible. From 1910 to 1915, Milton and Lyman Stewart paid for the publication of twelve pamphlets entitled *The Fundamentals of Christian Religion* (also called *The Fundamentals*). This series of pamphlets define what they and other like-minded people believe are the absolute, core beliefs of Christianity. Without the acceptance of these fundamental beliefs (the strict inerrancy of the Bible, the virginal conception and birth of Christ, Christ's substitutionary atonement for human sinfulness, the bodily Resurrection of Christ, and the Second Coming of Christ), these Christians argue, Christianity would be compromised. The movement's name, "fundamentalism," derives from the fundamental principles that these Christians strive to protect.[3]

Continued

2. *Dogmatic Constitution on Divine Revelation/Dei Verbum* (November 18, 1965), 3.11.
3. Ronald D. Witherup, *Biblical Fundamentalism* (Collegeville, MN: Liturgical Press, 2001), 6–10.

CHRISTIAN FUNDAMENTALISM ***Continued***

In their desire to protect these core beliefs, fundamentalists reject much of modernity, including methods for biblical interpretation that use tools of historical and literary analysis. One such methodology, known technically as the **historical-critical method**, employs a variety of tools, gleaned from such disciplines as history, literature, philology, sociology, anthropology, archaeology, geology, and physical science. By using these investigative tools, researchers try to answer questions about the text, such as, when and where was it written? Who wrote it? What is its literary form? What was the author's purpose in writing it? These researchers seek answers to such questions as a way of unlocking what the text meant originally and then interpret the text anew for current believers.[4] Because fundamentalist Christians reject modern literary and historical tools as a legitimate help to understanding the meaning of the biblical text, they would sharply disagree with the approach to biblical interpretation presented in this chapter.

To illustrate how the notion of **limited inerrancy** works when interpreting a biblical text through the lens of the historical-critical method, consider the following example.

In Genesis 30:25–43, Jacob outsmarts his father-in-law, Laban, who tries to cheat him of seven years' wages. Laban had agreed to pay Jacob all the "male goats that were striped and spotted, and all the female goats that were speckled and spotted, every one that had white on it, and every lamb that was black" (v. 35) that Jacob could find in Laban's herds. But Laban had his sons hide all such animals before Jacob could search the flocks. In response, Jacob whittled a few wooden sticks into various patterns of stripes and shapes. He then placed these sticks upright in the watering troughs so that the ewes and does in heat that came to drink would see these sticks as

4. Fundamentalists' rejection of the historical-critical method also includes all scholarly research on oral traditions, mythology, and folklore since the Enlightenment in the eighteenth century.

they mated. Jacob believed that the sight of these striped and speckled sticks would cause like markings to appear on the lambs and kids that would be born, and so it happened.

Modern science of course has disproved the ancient belief that *in utero* exposure to various sights affects the progeny of mating animals. The work in genetics begun by Gregory Mendel with pea plants in the nineteenth century and continued today on the molecular level by scientists in biogenetic labs has provided a much better understanding of the source of genetic traits. Does the variance between the biblical story and modern science make the Bible false? No, not at all.

Jacob's tale appears within a larger cycle of stories recounting God's fidelity to promises of land and offspring made to Abraham and his descendants (Gen 12:1–12; 15:1–21; 17:1–8). The truth that the modern reader may take from this tale is the very truth that its human author intended to make about God: God is faithful to the promises made to Abraham and his descendants. Because of the Divine's faithfulness to Abraham and his descendants in the past, modern readers may trust that God will be faithful to them. The Bible's human authors never intended its "truthfulness" to depend on its being an accurate portrayal of scientific, historical, geographical, or mathematical fact for every generation. The "truth" is in the story. If one looks only for fact or the accuracy of historical or scientific details, one can miss the real truth which, in the biblical stories, is the "truth" preserved "for the sake of [our] salvation."[5]

Discover for yourself

What truths does contemporary fiction tell about what it means to be a human being? Think of a few "truths" told by a novel, short story, or film you have read or seen. Jot down a list of five or six of these truths and be ready to explain why these are examples of "truth" even though they are found in fiction. Can you think of any biblical stories that relate truth in the same way as the examples you cited?

5. *Dogmatic Constitution*, 3.11.

A FAITHFUL AND RATIONAL READING OF THE BIBLE

All humans are shaped (and limited) by their historical, cultural, geographic, and linguistic milieu. The writers of the biblical texts were no different; they recalled their specific, local experiences of God. To find the meaning of these biblical stories, a twenty-first-century audience must grapple with how their original authors and audiences would have understood them. To tease out original meanings requires all the tools of literature, linguistics, history, archaeology, paleontology, sociology, and anthropology (i.e., the historical-critical method described above).

To read the Bible with understanding, one must build a bridge to the ancient text, the people who composed it, and those who first heard it. Then one can ask how the text might be understood today. But understanding tenth- through sixth-century BCE audiences and their experiences is not enough. One must also consider today's audiences and their experiences to clarify what a biblical text may mean for a modern community. If the Bible is the "living word of God" (i.e., inspired by God), it will act in a way that is descriptive of, or prescriptive for, believers' daily lives. Thus, the Bible must make sense to those who read it as a sacred text, as the Living Word. What it will mean for Christians and Jews today is related to what it meant for its first audience; what the text signifies to both communities is connected through its historical, literal meaning. It is only after establishing the "plain sense" or literal meaning of biblical texts that the modern reader can ask, "How might this text be understood in today's historical, political, social, and religious context?" These are the questions this chapter and chapter 4 will address.

ORAL TRADITION AND THE COMPOSITION OF THE BIBLE

Many books in the Bible, especially in the Old Testament, are compilations of shorter narratives. Many of these narratives began as campfire stories, as Israelite elders in their nomadic tribes shared their own and their forebears' experiences of God with the next generation. Such stories were gathered first orally, just as any family's memories, then written down by later generations who had heard these stories told and retold. Subsequent generations assembled these accounts into books. Still later, another familial group gathered up their ancestors' books that reflected their family's collective wisdom and memories about God. Finally, a last group of editors carefully organized these stories to best inform future generations of their family's collective memories and experiences of God. This is the process that the Israelites, or early Jews, and early Christians used to form the book known today as the Bible.

Preserving knowledge through memory and public recitation creates what is called an **oral tradition**. Sifting through oral stories, various written accounts, and books over the centuries to arrive at a final arrangement is the process of **canonization**. The final list of books on which the early communities came to agree is called the **canon**.

ORAL TRADITIONS

Oral traditions developed by nonliterate and preliterate societies preserved and passed on communal history, norms, and knowledge crucial to the survival of future generations. Members of such societies developed prodigious capacities to memorize, and so preserve, large amounts of detailed information. Often some members were provided with further training in this art. In Britain and Western Europe, some of these became bards to local kings. In ancient Greece, professional rhetors traveled from city to city and earned their living publicly performing epics such as Homer's *Iliad* and *Odyssey*.

IMAGE: © STAVCHANSKY YAKOV/SHUTTERSTOCK.COM

The Torah was preserved first through oral tradition, then in written form.

Continued

ORAL TRADITIONS ***Continued***

In ancient Ireland, the Brehons were entrusted with preserving and administering the Irish legal code. Ancient Israel had such specialists as well, and much of what we know today as the **Torah** or the **Pentateuch**[6] was preserved by them, first through an oral tradition passed down from one generation to another, and then, between the tenth and sixth centuries BCE, in written form.[7]

TWO STORIES, NOT ONE

The Bible contains two different stories that describe the creation of the world and humanity: Genesis 1 and Genesis 2–3.[8] It often comes as a surprise to readers new to biblical studies to find that there are two different accounts. Many have read Genesis 1–3 as one continuous story, with Genesis 1 as a broad overview of creation and Genesis 2–3 as a detailed version of the same story, similar to a scientist who increases the magnification of a microscope to capture more detail. Major problems, however, arise when a reader tries to reconcile the stories of Genesis 1 and 2–3 in this way.

Human Authorship of the Torah/Pentateuch

As noted above, Christians divide the Bible into two parts: the Old and the New Testaments. Christians further divide the Old Testament into four parts: Torah or Pentateuch, Prophets, Histories, and

6. The Hebrew word *Torah* is sometimes translated as "law" and at other times as "instruction." The Torah contains the first five books of the Bible: Genesis, Exodus, Leviticus, Numbers, and Deuteronomy. *Pentateuch* is a Greek word meaning "five scrolls"; it refers to the same five books.

7. The abbreviation *BCE* (Before the Common Era) refers to the same time period as the abbreviation *BC* (Before Christ).

8. Throughout this work, the terms *Genesis 1* and *Genesis 2–3* operate as abbreviations for the Genesis 1:1–2:4a and Genesis 2:4b–3:24 accounts respectively.

Discover for yourself

Open a Bible to Genesis 1, use a bookmark to hold your place, then open to Genesis 2–3 and place a bookmark there. Divide a piece of lined paper in half lengthwise. Label the left-hand side of the paper "Genesis 1" and the right-hand side "Genesis 2–3." Then answer the following questions for Genesis 1 and Genesis 2–3, writing your responses under the appropriate heading (give chapter and verse numbers to show where in the text you found each answer).

a. How many days does it take for God to complete creation?
b. When does God create the woman? When does God create the man?
c. From what does God create the woman? From what does God create the man?
d. What command does God give to the human beings?
e. What is God's name?

You have just completed an exercise in literary **criticism**, one of the methods used to identify what the biblical text meant to its original human authors and audiences. This exercise suggests that the creation stories were written by two different authors in two different historical periods to answer questions about faith and national identity raised by the daily circumstances of their respective audiences. The motivation to write each story arose, not from a desire to explain the scientific origins of the world, but to answer metaphysical questions about the meaning of human life, the purpose of human beings, and the nature of the world in which they live. For these reasons, and for other literary and linguistic reasons noted in this chapter's **exegesis**,[9] it is generally agreed that Genesis 1 and Genesis 2–3 had different authors who wrote for audiences widely distant in time and thus in social, political, and religious experiences.

9. The term *exegesis* refers to the act of "drawing out" the meaning of the text that its author intended and that its first audience experienced. The interpretation or meaning of any text will be linked to its original meaning, and its literary form (e.g., letter, poem, myth, law, and so on) will help to guide the interpreter in establishing the meaning its author intended.

Wisdom Literature. For both Christians and Jews, the Torah or Pentateuch forms the first section of the Bible. In this biblical context, the phrase "instruction for life" probably offers a better translation of the word *Torah* than the English word "law" because the five books, Genesis, Exodus, Leviticus, Numbers, and Deuteronomy, were used by the earliest Jewish audiences as a sort of instruction manual for how to maintain a balanced and just relationship with their Creator and one another. By the late nineteenth century, exercises such as the one you did in comparing Genesis 1 and Genesis 2–3 led scholars to accept the theory that more than one author composed the Torah.

Most modern scholars agree that Genesis 1 was written by a priestly collective,[10] most likely in Babylon, and that Genesis 2–3 was written much earlier by a "school" or collective, probably of royal scribes. The continuing debate about the sources of the Pentateuch will be set aside here. The focus of this chapter is the two "voices" or "authors" of Genesis 1 and Genesis 2–3, the "Yahwist" and the "Priestly" authors respectively, commonly referred to as "J" and "P."[11]

The Yahwist author (or "J") writes in folkloric style, creating warm and vivid portraits of people, often undergirding their lively stories with insights into their psyches. J draws an anthropomorphized image of the "Lord God," who actively engages human beings in intimate conversations and seems always ready to renew the relationship broken by the human partner. The priestly author ("P"), on the other hand, writes in a stolid, heavy voice, with measured tones. P's stories are rather formal, with God a distant and transcendent Divinity. P uses structure and repetition, at times giving his stories a stately, pageant-like feel. The priestly author may have been the final editor of the Torah in the postexilic period, though recent theories

---

10. The term *priestly* refers to the group of priests who accompanied the Israelites into exile in Babylon. This authorial collective wrote Genesis 1 and were the editors who placed it before Genesis 2–3.

11. The abbreviation *J* reflects the term's origin in late-nineteenth-century German biblical scholarship. Scholars employed the letter *J* as their abbreviation for the authorial source, who used the sacred name, Jhwh, to designate God throughout Genesis 2–3. English speakers will be more familiar with the transliteration for the Hebrew word which begins with *Y*, namely *Yhwh*.

about the Torah's composition cast doubt on P's final editorship. For our purposes, what matters is that Genesis 1 and Genesis 2–3 were written by two different groups of authors, widely separated in time and geography.

"COSMOGONY" AND THE ANCIENT NEAR EAST

In the Ancient Near East, people believed that the meaning and purpose of anything, whether a person, a nation, or an institution, could be found in its origins. In an entity's first moments, they believed, the intent of the Divine (or human) maker was sharpest. Consequently, to know the origin of something would give critical insight into its present meaning and purpose. The modern world retains something of this notion in the oft-heard adage, "The child is father to the man." Thus, for Israelites in the tenth through sixth centuries BCE, to understand the origin of the world was to grasp not only what God had desired for them and the cosmos in the first moments of creation, but also what God desired for them now. So, a story could be developed describing the dawn of human time, either in great detail or with a majestic sweep of the authors' literary brush, without any intention of rendering a scientific account. The Israelites created just such a story for themselves, twice: first with Genesis 2–3 and again with Genesis 1.

Like other peoples of the Ancient Near East, the Israelites developed their own **cosmogony**, or "story [about] the origin and development of the universe,"[12] to articulate who they were, who God was, why they existed, and what they could expect from God, one another, and the world around them. They recombined Mesopotamian and Canaanite creation stories, events, and characters, which they had learned from centuries of interaction with these neighboring cultures. In doing so, the Israelites created a new cosmogony that stood in stark contrast to the worldview promoted by worshippers of Baal, Astarte, Marduk, Tiamat, and many other, lesser-known gods

---

12. *Merriam Webster's Collegiate Dictionary*, 10th ed., at *http://dictionary.reference.com/browse/cosmogony*.

and goddesses of the Ancient Near East.[13] As the Israelites borrowed crucial pieces of their neighbors' creation accounts, they removed ideas and events that did not reflect their own experiences or expectations of God and the world and added others that did.

As noted above, Genesis 1 and Genesis 2–3 were written in widely separate historical periods and are distinctly dissimilar in style and tone. Yet they reflect similar understandings of who God is, who human beings are, and what these human beings can expect from God, one another, and the rest of creation. These accounts challenged the rest of ancient humanity's presumptions about what it means to be a human being in a world of chaos and violence.

GENESIS 2–3: THE YAHWIST ACCOUNT

By the beginning of the tenth century, Israel had become a nation of twelve tribes, united under a new, powerful king, David. While David consolidated his power and secured the throne, the two superpowers of the age (Egypt and Assyria) were experiencing major internal strife and had not the time, the military force, or the attention to spare for dealing with a small, upstart, backwater entity like Israel.

Israel had begun to develop a stable food source through farming, and the people's lives and livelihoods were protected by the power of their new king. Less time had to be spent eking out a living by nomadic herding or protecting flocks from itinerant human and animal marauders. Permanent settlements of Israelites began to grow into small cities, and David made Jerusalem (a large city that had been captured by the Israelites) the nation's religious and political capital. Israelites finally had the luxury of devoting more people to work in areas other than food production and physical protection, and they had leisure. The new circumstances allowed the united kingdoms of Israel and Judah to refine their religious, social, civic, and cultural practices in ways associated with newly urban societies.

---

13. See Richard J. Clifford and John J. Collins, eds., *Creation in the Biblical Traditions,* Catholic Biblical Quarterly Manuscript Series (CBQMS) 24 (Washington, DC: Catholic Biblical Association of America, 1992), and Richard J. Clifford, *Creation Accounts in the Ancient Near East and in the Bible,* CBQMS 26 (Washington, DC: Catholic Biblical Association of America, 1994), passim.

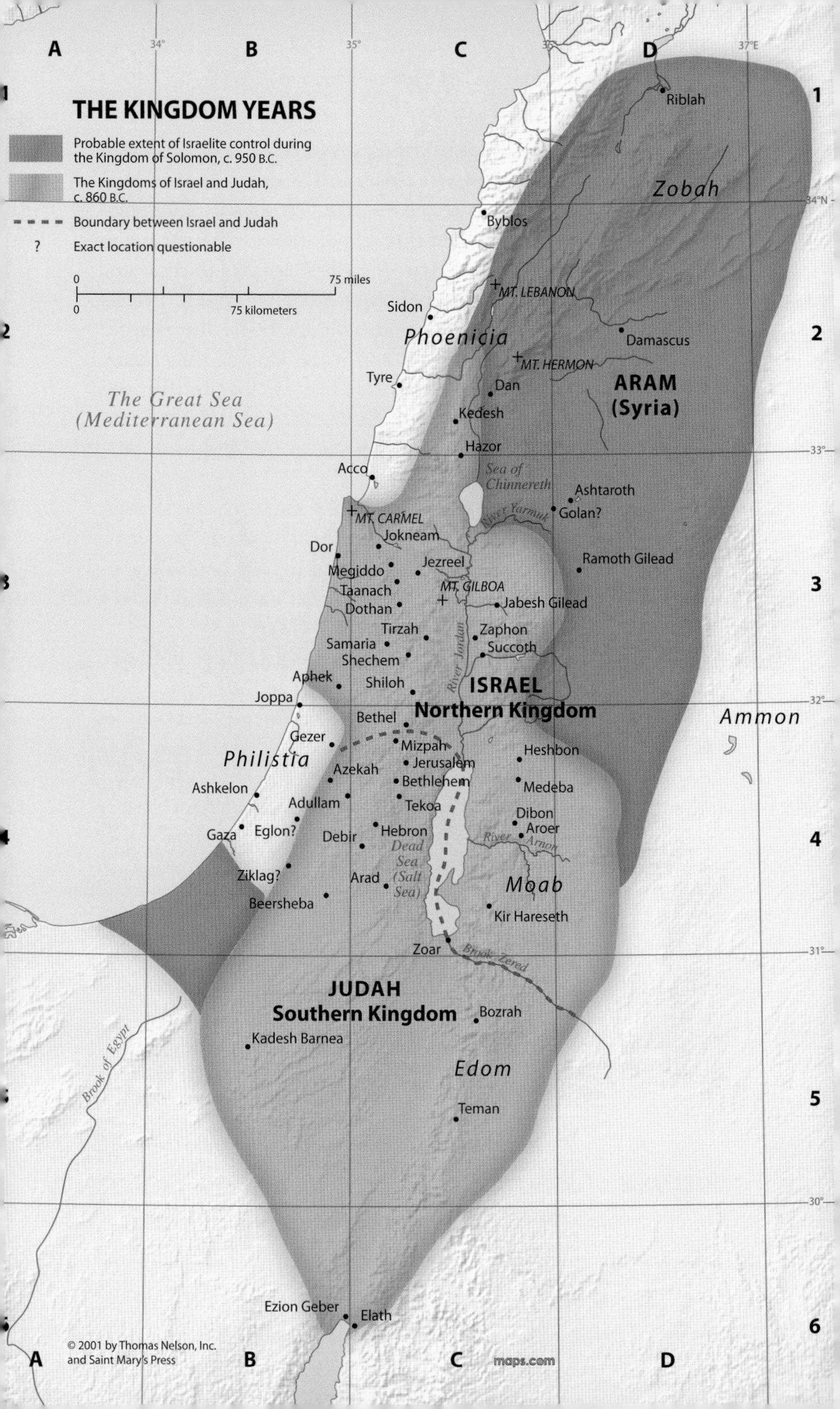
THE KINGDOM YEARS
Probable extent of Israelite control during the Kingdom of Solomon, c. 950 B.C.
The Kingdoms of Israel and Judah, c. 860 B.C.
Boundary between Israel and Judah
? Exact location questionable
0
75 miles
0
75 kilometers
A
B
C
D
34°
35°
36°
37°E
34°N
33°
32°
31°
30°
1
2
3
4
5
6
Riblah
Zobah
Byblos
MT. LEBANON
Sidon
Damascus
Phoenicia
MT. HERMON
Tyre
Dan
ARAM (Syria)
The Great Sea (Mediterranean Sea)
Kedesh
Hazor
Acco
Sea of Chinnereth
Ashtaroth
Golan?
River Yarmuk
MT. CARMEL
Jokneam
Dor
Ramoth Gilead
Megiddo
Jezreel
MT. GILBOA
Taanach
Jabesh Gilead
Dothan
Tirzah
Zaphon
Samaria
Succoth
Shechem
River Jordan
Aphek
Shiloh
ISRAEL Northern Kingdom
Joppa
Bethel
Ammon
Gezer
Mizpah
Heshbon
Philistia
Jerusalem
Azekah
Bethlehem
Medeba
Ashkelon
Adullam
Tekoa
Dibon
Gaza
Eglon?
Hebron
Aroer
Debir
Dead Sea (Salt Sea)
River Arnon
Ziklag?
Arad
Moab
Beersheba
Kir Hareseth
Zoar
Brook Zered
JUDAH Southern Kingdom
Bozrah
Kadesh Barnea
Brook of Egypt
Edom
Teman
Ezion Geber
Elath
maps.com

King David (1000–960 BCE) established a school of scribes, modeled on schools already well established in other Ancient Near Eastern nations such as Egypt and Assyria. David's son, King Solomon (960–922 BCE), enlarged this school, which was later maintained by his successors. This corps of scribes provided the king and his royal household with a kind of ancient internal revenue service or inventory control. The king, and later his heirs, also charged these scribes with writing the history of the king and his family. The scribes had several sources for this history, but the chief one would have been the oral tradition inherited from their wandering Israelite forebears.

Aside from Israelite oral traditions, the royal clerks would also have drawn from stories and literary forms found in neighboring, non-Israelite cultures—Phoenician, Canaanite, Akkadian, Sumerian, and Egyptian—which they had encountered in their nomadic years and in ongoing trading ventures. These scribes were operating on a precept still familiar today: why reinvent the wheel?

First, the scribes would have had to translate these borrowed stories from Canaanite (Ugaritic), Akkadian, and other languages to Hebrew. Later, they would adapt and interweave these stories with their own ancestral stories. The new stories would have been designed not only to support and lionize the kings (and their ancestors and royal descendants) who paid for the work but also to chastise and admonish the people for their failings. In addition, these stories would have addressed questions of identity, meaning, and purpose in Israelite society. Such questions would have arisen within the people because of the massive changes they had experienced in their social, familial, and religious structures and practices over their last couple of hundred years. What better way for the Israelites to answer such primal questions than to devise a story of the origins of their world? In responding to these questions, the scribes would reveal how their God had worked, and continued to work, for the preservation of the people of Israel.

Analysis of the Text

We begin with the older of the two creation stories found in Genesis 2:4b–3:24. As noted above, scholars believe that the corps of royal scribes first began to write under Kings David and Solomon. Their successors in the royal civil service continued to write for the heirs of the Davidic dynasty even after the twelve tribes had split into two separate kingdoms. By the 600s BCE (some three hundred years after David organized the first school of scribes and at least seventy-five years after the fall of the **northern kingdom**), these scribes had become the preservers of earlier written traditions, the oldest likely dating from the kingships of David and Solomon.

The Yahwist or J writer offers a creation story with two distinct parts: (1) God's creation of the garden and humankind (2:4b–25) and (2) a demonstration of human frailty in the face of temptation and Divine compassion and fidelity in the face of human perfidy (3:1–24). The writer draws on several Mesopotamian creation myths (i.e., the ***Enuma Elish***, the *Adapa* legend, and the epics of *Atrahasis* and *Gilgamesh*), as well as from images in Israel's store of Canaanite myths.[14] Adapting the creation accounts the Israelites had learned from trade and intermarriage with their Canaanite and Mesopotamian neighbors, the J author addresses basic questions of meaning that had arisen among the members of the changing and frequently threatened Israelite society. He employs this first story to set the stage for the long, primeval history narrative in Genesis 2–11. This creation narrative (Gen 2–3) then becomes the explanatory backdrop for why humanity so desperately needs the covenant that God would first make with Abram (Gen 12), and by extension with his wife, Sarai, and so with all their descendants.[15]

Beginning the story with the creation of the garden and humankind, J says that on the day the Lord God makes the earth, it is barren of plant or animal life, and because the Lord God has not yet made

---

14. The word *Mesopotamian* refers to the land that lies between two great rivers, the Tigris and the Euphrates. The nations most associated with Mesopotamia are Sumer, Akkad, Assyria, and Babylon. From 2200–500 BCE, these nations, and some others that spoke the Semitic languages of the eastern Mediterranean coast (e.g., Ugaritic, Phoenician, Moabite, Ammonite, and Hebrew, a collection of languages sometimes termed *Canaanite*), contributed many images and details from their cosmogonies to the J writer's project, an account that would reflect the Israelite view of God, human beings, and the world.

15. Claus Westermann, *Genesis 1–11*, Continental Commentary (CC) (Minneapolis, MN: Fortress Press, 1994), 194.

it rain, there is no "earth creature" to work the "earth" (vv. 4b–5).[16] Into this desert, reminiscent of the dry conditions of the Israelites' land, the Lord God causes a mist to go up from the land and to water the whole face of the earth (v. 6). Land covered with such a mist would have seemed like paradise to a people used to constant water shortages. After creating the watery conditions under which human life can flourish, the Lord God fashions the human species (*'adam*) of dirt from the earth (*'adamah*) (v. 7). The J author is playing on the Hebrew words for "human species" and "earth" or "ground," using their similar sound to connect them as intimately as humanity is connected to the land and its sustaining fruit.

This image of a god creating a human being from the dirt of the earth frequently appears in the mythology of early cultures, as in the Mesopotamian epic of *Gilgamesh* in which a goddess creates a man from a piece of clay.[17] In contrast to these other accounts, this image of God in Genesis 2 is connected with the later story of Israelite origin, where, after God's initial care for creation and intimacy with the first humans (Gen 2–3), Abraham and his descendants enjoy privileged interactions with God (Gen 12–50).

The rupture that this human couple introduces between themselves and God (3:1–24) does not dissolve God's relationship with them or with their immediate offspring, as the primeval history shows (Gen 4–11).[18] In fact, the stories in the rest of Torah show God actively engaged in repairing this breach through establishing covenants with Noah, Abraham, and Moses.

In v. 7b God breathes into the nostrils of this newly created earth creature "the breath of life"; the inanimate object fashioned of "dust of the ground" becomes "a living being," not a body with a separable soul, but a living entity whose being is enlivened by the breath of God. For this human being, God then plants "a garden in Eden" (v. 8), filling it with plants that are both beautiful and nourishing (v. 9). Yet the creator also plants "the tree of life" and "the tree of the knowledge of good and evil" (v. 9).

---

16. See Phyllis Tribble, *God and the Rhetoric of Sexuality*, Overtures to Biblical Theology (Minneapolis, MN: Fortress Press, 1978), 77.

17. James B. Pritchard, *Ancient Near Eastern Texts Relating to the Old Testament with Supplement* (Princeton, NJ: Princeton University Press, 1969), 74.

18. Westermann, *Genesis 1–11*, 195–96.

The tree of life appears again only at the end of the narrative (3:22–24); it frames both the drama about to unfold (2:15–3:24) and the second tree mentioned, the tree of the knowledge of good and evil. In 3:22 the Lord God reflects, "Behold, the man has become like one of us, knowing good and evil; and now, [he must not be allowed to] put forth his hand and take also from the tree of life, and eat, and live forever." There is but one tree in the story, which the narrator describes in two ways: the tree of life and of the knowledge of good and evil.[19] This one tree represents the wisdom that is the sole province of God and that human beings may not grasp (cf. Jb 15:7–9, 40; Prv 30:1–4).[20] The human couple's attempt in Genesis 3 to obtain such wisdom by eating the forbidden fruit (2:17) forms the center of this two-part story. The couple's choice to eat from the forbidden tree and God's response exemplify a key aspect of what the narrator believes it means to be a human being: a human being is not God.

The four verses that interrupt the flow of the story (vv. 10–14) remind readers that the blessing of fertility is brought to earth by "a river [that] flows out of Eden" and "divides and becomes four branches," a blessing that finds its source in God. The Lord God of Genesis 2–3 is the only source of fertility for the land; making offerings to the gods of Israel's neighbors or overlords will not ensure the fertility of the land or the people. These verses remind and subtly warn the people, especially the kings, that they are not to find the blessing of fertility in non-Israelite religious practices.[21]

J resumes the main narrative in v. 15, reminding readers that God has put the human being in the Garden of Eden. J then identifies the human being by its relationship to the garden: this is the one who is "to work it and keep it" (v. 15). With these words, the J author shows a human being who reflects the image of God in vv. 4–9. God has cared for the earth on a grand scale; like God, the human being is to care for the earth, but on a creature's scale. In contrast to the *Enuma*

19. Westermann, *Genesis 1–11*, 212.

20. Gordon J. Wenham, *Genesis 1–15* (Word Biblical Commentary [WBC]; Waco, TX: Word Books, 1987), 63.

21. The scribes who wrote and edited Genesis 2–3 believed all of the kings of Israel and all but three of the kings of Judah (David, Hezekiah, and Josiah) had failed as kings because they had worshipped and sacrificed to gods other than Yhwh. See 1 and 2 Samuel and 1 and 2 Kings for their stories.

Elish (6:33–36) and the *Atrahasis* epic (1:190–97), the purpose of the human being is not to labor in place of the gods, but rather to imitate God's care and creativity.[22]

In v. 16 the story returns to the trees described as "good for food" (v. 9), with the Lord God giving the creature unrestrained access to eat their fruit with one exception (v. 17): the new creature must not eat from the tree of the knowledge of good and evil. Neither God nor the narrator justifies the prohibition; it simply is. The story of chapter 3 will unfold around this tree, the command forbidding humanity its fruit, and the serpent who questions the Lord God's command.

In the next section of his creation story, the J writer portrays the Lord God's continued attentiveness to the new creature's basic needs. Noticing that food and work are insufficient to sustain the human creature, the Lord God says, "I will make him a helper fit for him" (v. 18). The Hebrew phrase literally and awkwardly means "a helper like near to him."[23] The J writer appears to argue that human beings' need for others "like near to" them (that is, human community) has been "hard wired" into them; it forms a constitutive element of humanity's psychological and emotional makeup.[24]

The first creatures God forms as companions for the man are the animals, but God does not breathe into them the Divine breath. Nonetheless, the narrator calls all of the animals and birds "living creatures," as he had called the human creature in v. 7. By having the human creature name the animals and birds, the narrator signals that they are not things, but living beings with agency.[25] Yet none of these "living creatures" provides a "helper like near to" the human creature,

---

22. Wenham, *Genesis 1–15*, 67.

23. David W. Cotter, *Genesis*, Berith Olam (BO) (Collegeville, MN: Liturgical Press, 2003), 31. The Hebrew word *ʿēzer*, translated as "helper," means aid that comes from God alone, particularly when facing deadly peril (cf. Ex 18:4; Dt 33:7, 26, 29; Hos 13:9; Ps 20:3 et al.). Because God is often the helper in these other passages, the "helper," of v. 18 should not be seen as subservient to the man. In fact, the consonantal root of the word *help* in Hebrew (*ʿoz*) is also the root of the Hebrew word for *strength*. Cotter suggests that J is playing with the two like-sounding words: *ʿēzer* for "help" and *ʿoz* for "strength." He infers that J intends the woman to be the "strength" that helps the man, just as Yhwh's "strength" in battle helps the man (32). He argues also that the Lord God has formed the human being so that he will need the help of another creature "like near to him" with whom he can establish a mutual relationship.

24. Westermann, *Genesis 1–11*, 227.

25. Ibid., 228.

whose lonely state the Lord God recognizes as "not good" (v.18). So the Lord God fashions a "helper like near to" the human creature (v. 21) who, unlike the animals, will prove "fit" for a relationship of intimate human community.[26]

The woman fashioned by God building up a rib (vv. 21–22) recalls the human figurines made of clay molded onto sticks that were unearthed in digs at Jericho.[27] This semblance suggests that behind these verses rests an old tradition about the construction of humanity, which J has incorporated into his narrative. In addition, J's use of the word "rib" may recall the Sumerian word for "rib," which also means "life." J creates a wordplay in which the "lady of the rib" is also the "lady of life"; he also anticipates the woman's proper name, which the man will give her in 3:20. That name, Eve or *Ḥavvâ*, is a form of the Hebrew word for "life" and will provide the woman's epithet, "mother of all life" (3:20).[28]

The Lord God then takes the woman to the man (v. 22), who immediately sees that she is the "helper fit for him," and the man poetically proclaims both his recognition of her identity and his acceptance of her (v. 23). The empty solitude in which he has lived has been banished. The J author, in addition to having the man praise the woman, continues to play with similar sounding words to underscore the connections being drawn: "She shall be called *a* 'woman' [*'îshah*], for she was taken out of *a* 'man' [*'îsh*]" (v. 23). On one level, this verse provides more evidence for J's etiology[29] of marriage in v. 24; yet on another level, this verse recalls a genuine human experience of encountering another human being with whom to share intimacy and a sense of human community.[30]

The next verse, "the man and his woman were both naked, and were not ashamed (v. 25)," provides a bridge between J's story of creation and the subsequent story of humanity's alienation from God and from one another. It signals that while there was a time when the

---

26. Ibid., 230.

27. R. Amiran, "Myths of Creation of Man and the Jericho Statues," *Bulletin of the American School of Oriental Research* (*BASOR*) 167 (1962): 23–25.

28. Westermann, *Genesis 1–11*, 230.

29. An etiology is a folk explanation for a physical, cultural, or scientific phenomenon. The Bible contains many etiologies.

30. Cotter, *Genesis*, 33.

man and the woman were not ashamed, they both soon would be. The couple's lack of shame in their nakedness symbolizes the state of their relationship with God and one another—open, unself-conscious, intimate.[31] The absence of shame does not, as some commentators have suggested, indicate that the couple has not yet discovered sexual intimacy or sensuality.[32] Rather, the story that follows provides an etiology for the disruption humanity experiences in its relationship with God and with one another, and not a description of the first sin committed by a historical person. Just as Genesis 2:4b–25 is not an account of two historical persons interacting with God at the creation of the earth, neither is Genesis 3:1–24. In a broad sense, like chapter 2, discussed earlier in this chapter, and chapter 1, to be discussed, the literary form or genre of Genesis 3 can be called **myth.**[33]

Genesis 2:25–3:7 contains a brief picture of human life before alienation (2:25), the process of that alienation (3:1–6), and the initial effect of that alienation on human life (3:7). J's wordplay with like-sounding words—*ʿarummîm* or "naked" (2:25) and *ʿarum* for the serpent's "cunning" (3:1)—more closely connects the serpent's "cunning" and the man and the woman's nakedness. When the word "naked" appears in 3:7, we meet the fruition of the subtle hint and the reversal of the openness, intimacy, and unself-consciousness alluded to in 2:25's phrase, "both naked and . . . not ashamed." The adjective "naked" frames the material between 2:25 and 3:7 and creates a self-contained literary unit.

In the scene between the woman and the serpent, the serpent is pivotal for moving the story's plot forward. The serpent, described as "the most cunning " of all animals (3:1–5) (a characterization shared

---

31. Ibid.

32. See Westermann's discussion of this misinterpretation that has been used to claim that the "fall" from grace by the first couple was caused by some sexual sin (*Genesis 1–11*, 233).

33. The word *myth*, when used in its oldest and fullest sense, points to a story whose task is to present the deepest truths of what it means to be a human being. When everyday language becomes inadequate to express the most profound realities of beauty or goodness, evil or corruption, writers and poets turn to metaphor as a way to reach beyond the limits of daily speech. Their intent is to evoke in their audience's subconscious a visceral recognition of the "meta" truth that the metaphor-story reveals about humanity, its God, and the created world. That is the lens through which the author of this chapter will proceed.

with the mythology and common wisdom of surrounding Near Eastern cultures), also serves as a foil against Ancient Near Eastern beliefs about the power of serpents in the face of Yhwh God. Popular belief in the Ancient Near East attributed to the serpent proverbial wisdom and cunning (cf. Mt 10:16) and the power to rejuvenate (shedding its skin).[34] It had become a symbol of immortality (e.g., Ez 29:3) as well as of death.

Indeed, in Ancient Near Eastern mythology, the most fearsome creature of all was the large, snakelike sea monster (cf. Is 27:1; Jb 26:18) that inhabited the unknown depths of all bodies of water. The J author removes this monster from the great seas and rivers, shrinks its size, and brings the fearful creature into the light of day. In so doing, J destroys the power that the Israelites may still ascribe to the serpent as a local deity. J makes the serpent ordinary and places it under the aegis of God's creation by designating it as "the [most cunning] of all the wild animals which Yhwh God had made" (3:1).[35] By continuing the play on the words *naked* and *cunning* first encountered in 2:25 and 3:1, the writer depicts a couple who will seek to be as cunning as the serpent in v. 1, only to discover that they have become naked in that effort (vv. 7, 10), "snake-naked" in fact.[36]

The serpent's cunning becomes evident when it broaches the topic of Yhwh God's command not to eat from the tree at the garden's center (3:1; cf. 2:9).[37] In the dialogue between the woman and the serpent, the author suggests the man's silent presence and complicity with the woman by having both speakers use the plural "you" and "your." The serpent begins to subvert the couple with a seemingly innocent but cunningly calculated question, as though seeking assurance that its information about God's prohibition is correct (3:1). It has changed Yhwh God's "command" in 2:16 to a mere "say" in 3:1, weakening its force in the minds of the woman and man. And it alters the identification of God by dropping the sacred name, Yhwh.

---

34. Nahum M. Sarna, *Understanding Genesis* (New York: Schocken Books, 1970), 26.

35. U. Cassuto, *A Commentary on the Book of Genesis*, trans. I. Abrahams (Jerusalem: Magnes, 1964), 141.

36. Cotter, *Genesis*, 34.

37. Nahum M. Sarna, *The JPS Commentary: Genesis* (Philadelphia/New York: Jewish Publication Society [JPS], 1989), 24.

The serpent also widens God's prohibition to a ban on fruit from all trees in the garden, rather than just the tree at the center. The woman hastens to God's defense, correcting the serpent's harsh presentation of what God has "said." The serpent's trap, baited with a distorted image of God, successfully lures the woman into dialogue to defend the Creator. In attempting to defend Yhwh God, the woman joins the serpent's misrepresentation of God in three ways: (1) she adopts the serpent's change of God's name herself (v. 3), (2) she repeats the serpent's change of God's direct speech from the strong "command" to the weaker "said," and (3) she adds a clause that God did not say, "nor shall you touch it" (v. 3). The J author signals with each change of the Divine's speech that the woman and the man have moved further from God and closer to the serpent and the values it symbolizes.[38]

To ask why the serpent approaches the woman rather than the man is not sexist. This question reflects a twenty-first century concern that results from a long history of textual misinterpretation. What the J tradition can answer is that it sees the man and the woman relating to one another as equals (viz., *'adam* as humanity and *havvah* as life, 3:20).[39]

With the couple's seduction well under way, the serpent replies to the woman's final concern about dying (v. 3) by contradicting Yhwh God's pronouncement in 2:17. The serpent declaims, "You will not die" (3:4). This statement is a lie, believable to the woman and the man because it contains vestiges of truth (they do not "die"), a shred the serpent will use again later to sow doubt in Genesis 3:5. Neither the serpent nor the woman in this dialogue names God using God's personal name, Yhwh, although the phrase "Yhwh God" is used throughout the rest of Genesis 2–3 by J. "Yhwh God" is personal, and its use indicates a relationship of intimacy between creature and creator, a covenant partnership (cf. Ex 9:30; 2 Sm 7:25; Pss 72:18, 84:12). Neither the serpent nor the woman can name God using a name so contrary to the restrictive and parsimonious character that the serpent has created in the woman's mind. When the woman replies to the serpent using the name "God" without the prefaced

38. Westermann, *Genesis 1–11*, 239.

39. D. E. Gowan, *From Eden to Babel* (International Theological Commentary; Grand Rapids, MI: Eerdmans, 1988), 53.

"Yhwh," the J tradition recognizes the seduction taking place and hints at the disobedience to God soon to follow.[40]

The serpent continues its corruption of the couple with two additional lies: "your eyes will be opened" and "you will be like God, knowing good and evil" (3:5). Their "eyes will be opened," but to their nakedness, and they will become "like God, knowing good and evil," but at a cost the serpent fails to mention.

This phrase in v. 5 expresses the desire to make judgments about the conduct of one's life, to choose between good and evil.[41] The couple desires to conduct their life independently of the Creator's guiding wisdom. This drive for autonomy is at the center of the couple's surrender to the serpent's cunning.[42] They are successful in becoming "like God" and in asserting their autonomy, but they do so at a cost. They become alienated in their relationship with their Creator and with one another, as the remainder of Genesis 3 demonstrates. The serpent lured the couple into betraying God's command by planting seeds of doubt about God's motives in restricting them from the fruit of one tree. The couple's urge to exercise human knowledge independently of God's wisdom springs from their failure to trust God's motives in restricting them from wisdom, and so both eat the fruit (v. 6). By eating the fruit, they become "like God, knowing good and evil" (v. 5), but do not die. The couple lack faculty to exercise such knowledge wisely, however, and they sin.

From this narrative, readers might conclude that the source of human sin is often found in humanity's determination to make decisions beyond its capacity. Genesis 3 indicates that the J author recognized the human inclination to act unwisely; humans fail to trust the Goodness that created them. This story describes a process common to every human being—sin—and ascribes its cause to human beings who unwisely choose to distrust Goodness. The effect of their defiance is immediate and ironic. Their eyes open and they acquire new knowledge: they are naked, but do not speak, and they separate from one another to find material to cover their nakedness.

---

40. Wenham, *Genesis 1–15*, 57.

41. Please note, as mentioned above, the phrase "to know good and evil" does not refer to a before-and-after state of sexual knowledge.

42. Sarna, *JPS*, 19.

Speech, which first marked the man's recognition of his communion with the woman (2:23), has disappeared. They hide, at first from one another in their search for clothing (v. 7), and then from Yhwh God who still seeks them out (v. 8).[43] Their realization that they are naked, the frenzied search for something to cover themselves, and their fear of God signal the transformation in their relationships with one another and with God.[44]

Human speech returns when God returns, but the relationships of intimacy and trust do not (vv. 8–13). The woman's and the man's rush to avoid responsibility for their choices would be almost comical if it was not so pathetic (vv. 11–13). The man even tries to blame God for his choice to eat the fruit because God had created the woman, and the woman tries to blame the snake. The previous intimacy and vulnerability the human couple had shared with one another and with God has been replaced by a poor imitation of the serpent's cunning.[45]

A series of punishments, written as poetry, follow for the three miscreants in this drama (vv. 14–19).[46] Each punishment is an etiology (see n. 29) explaining a common experience in daily life among the people of ancient Israel. The punishments explain why the snake crawls on the ground and all humanity and animals fear it (vv. 14–15), why women experience great pain during childbirth and live lives dominated by men (v. 16), why men suffer a life of toil and sweat, and why human beings eventually die (vv. 17–19). On one level the punishments are simply descriptive, accounting for observations and cultural practices among the Israelites; on another level, however, the punishments serve as warnings from the J author to his audience: wrongdoing has consequences, and do not participate in foreign cults.

First, the snake was a potent symbol of various Canaanite worship. Yhwh God demonstrates complete authority over the snake, and so over the Canaanite cults it represents, by decreeing a

43. Cotter, *Genesis*, 35.

44. Westermann, *Genesis 1–11*, 254.

45. Cotter, *Genesis*, 35.

46. When poetry is found embedded in prose, it often signals that the poetry was composed orally at a time much earlier than the composition of the surrounding prose.

miserable future, and warns the Israelites to avoid interaction with the Canaanite cults by announcing perpetual hostility between the serpent's offspring and all humanity (vv. 14–15). Yhwh God then admonishes the Israelite kings, who had consistently married daughters of foreign kings in order to "seal" political treaties with neighboring powers. (When a new royal wife would arrive in Jerusalem, she would bring with her various images of her national or tribal gods, including the snake, as well as priests to perform appropriate sacrifices.) In v. 16d, Yhwh God warns all Israelite men who had taken foreign wives to "rule over" their wives' pagan practices. Men were to exercise the very "cunning" (another way of saying "wisdom") for which the serpent was a symbol in the Ancient Near East, but for which Yhwh God alone was the true source, and suppress the foreign cults practiced by such wives lest they also become ensnared by them.[47] In its historical setting, this admonition did not refer to men claiming authority over women, nor should it be read that way today.

The last punishment (vv. 17–19), in addition to its etiological function accounting for the harsh reality of agricultural work, may also preserve part of an old critique of the Israelites' adoption of neighbors' farming practices. When the Israelites first entered the land of Canaan (ca. 1200 BCE), they were in transition from nomadic herders of goats and sheep to farmers who were settled permanently on land. They would have had to learn how to farm from their Canaanite neighbors, whose agricultural methods included various practices for propitiating the deities of the land and water. The Israelites would have learned to engage in those polytheistic rituals of their neighbors right along with employing the skills necessary for the actual cultivation of crops. It is likely that this last punishment underscores the first two, by warning the Israelites a third time against worshipping foreign gods and participating in fertility cults. To be an Israelite, the story advises, is to be a human being who depends solely on Yhwh God and does not take out an "insurance policy" with neighbors' gods in case Yhwh God fails to live up to their trust.

47. See 1 Kings 11 for a description of this phenomenon by a writer who was probably contemporary to the later editors in the J tradition.

J returns to prose in 3:20, when he has the man (*'îsh*) name the woman again, as he had done in 2:23; this time, however, instead of giving the woman a derivative of his name (*'îshshah*), the man gives her a new name that signals their changed reality. They have become alienated from one another and from Yhwh God. The intimacy they once shared with one another, which the word pair (*'îsh/'îshshah*) revealed, is now gone. Their lives have changed, yet are not over, as the woman's new name intimates. The man calls her *Ḥavvâ*, "living woman" or "woman alive," "because she [will become] the mother of all life" (v. 20).

Neither man nor woman has met with the death the reader thought Yhwh God had asserted would be the result of their eating fruit from the tree of the knowledge of good and evil (2:17). But their active mistrust of Yhwh God in 3:5–6 does kill the basis for the man's cry of joy in 2:23. Their once-intimate union with one another and with Yhwh God has become alienation, a form of emotional death. This deadly alienation will pursue humanity throughout the rest of the Genesis narrative, and God will spend the rest of biblical history repairing its effects. Although a form of death did come to the couple, Yhwh God did not abandon them to it, as seen in the deity's providing them with protective clothing to cover their nakedness (3:21).

In the last three verses (vv. 22–24), J returns to the tree of life last mentioned in 2:9. As he did in 2:9, J entwines his comments about the tree of life with the motif of knowledge of good and evil, found also in 3:5. The tree of life is the tree of the knowledge of good and evil. The human couple have sinned by eating from the tree forbidden to them, and they have become alienated from the life they had lived in Yhwh God's park-like garden. They are expelled into another kind of life (see the rest of the J primeval history account in Gen 4–11), where their relationship with Yhwh God and one another will continue to be affected by their sin: their failure to trust God and their grab for knowledge that God alone has the maturity to wield. The human couple tries to return to the garden and the life they had known before their encounter with the serpent stripped them of their trust in Yhwh God. But cherubim bar them from entering. In the Mesopotamian myth tradition, cherubim are fantastic creatures whose pictures adorn the walls of the Temple (cf. Ex 26:31; 1 Kgs 6:29); a molded pair even stand as the throne for the ark of God's presence

IMAGE: WIKIMEDIA, PUBLIC DOMAIN

In Genesis, cherubim bar the human couple from reentering the garden. In the Mesopotamian myth tradition, cherubim are fantastic creatures like this human-headed winged bull bas-relief found in King Sargon II's palace at Dur Sharrukin in Assyria (now Khorsabad in Iraq), c. 713–716 BC.

(Ex 25:18–22).[48] In Ezekiel 9:3; 10:1–22 and Psalms 80:1, the cherubim also serve as Yhwh's throne.[49] The revolving, flaming sword that accompanies the cherubim also prevents the couple's reentry into the garden; it symbolizes the flashing lightning that Israelites viewed as God's anger striking (cf. Jer 46:10; Is 34:5; Zep 2:12).[50]

DISASTER: THE BABYLONIAN EXILE

The creation account presented in Genesis 1 comes from one of the most wretched periods in Israel's history. In the waning years of the seventh century, long after the Assyrians had destroyed the northern

48. Wenham, *Genesis 1–15*, 86.
49. Ibid., 274.
50. Ibid., 275.

kingdom of Israel and forcibly moved the surviving Israelites to the far regions of the Assyrian Empire (722 BCE; see 2 Kgs 17:6),[51] the Kingdom of Judah fell under siege by the newest superpower, the Babylonian Empire, which was ruled by King Nebuchadnezzar.

Successive Davidic kings and their advisors had made a series of disastrous alliances and political missteps that reduced the **southern kingdom** of Judah to the city of Jerusalem and a small circle of land around it. For the second time in twelve years (first in 598 and next in 586), Nebuchadnezzar and the Babylonian army laid siege to Jerusalem, with no interest in negotiating. The Babylonian army encircled Jerusalem for two years, waiting for the inhabitants to surrender to avoid inevitable famine. When Judah's king, Zedekiah, and some retainers were caught trying to flee Jerusalem by night, Nebuchadnezzar and his men descended on the city, demolishing its walls and razing Solomon's Temple.

---

51. Unless otherwise indicated, all dates in this chapter are BCE.

The northern population of Israelites, usually associated with ten of the twelve fabled tribes of Israel, broke away from the Davidic dynasty-ruled southern kingdom (Judah) in 922 and formed their own kingdom. Without a divinely appointed royal dynasty from which to draw their kings, the northern kingdom (Israel) found itself suffering from frequent wars of succession for the throne. Included as a part of such wars, which usually occurred between the offspring of wealthy and powerful families, were incursions into Israelite territory by foreign powers, from whom the various warring Israelite factions had sought military aid. The cost of doing such business with foreign nations almost always required that the new, northern king acknowledge the foreign nation's king as his overlord, cede part of his territory to his "ally," and pay a "war tax" for the foreign nation's help in overthrowing the previous Israelite king. Usually these "treaties" were sealed through an exchange of royal daughters, probably from one of the many concubines both kings retained in addition to their wives. By marrying or formally accepting as a royal concubine the other king's daughter, the kings were signaling that these two "houses" or "dynasties" were firmly tied to one another by the exchange of blood.

The problem for the Israelites (and the Judahites in the South as well) was that such "marrying out" by the king meant that he and at least some of his people would have to support and engage in Gentile worship practices. Their "two-timing" of the Lord God with the gods of the Gentiles did not go unnoticed by some prophets and scribes, the latter of whom recorded strong critiques of such kings and their wives in the royal chronicles (see 1 Kgs 11 and 18), often shaping their narratives to demonstrate that this infidelity to the God of Israel (or Judah) had brought devastating consequences to the king and people. The overarching theme of "disobey, and be punished; obey, and be blessed," which is found in all of the Deuteronomistic Histories (Jo, Jgs, 1 and 2 Sm, 1 and 2 Kgs), as well as Deuteronomy itself, reveals the probable mind-set of the final shapers of Genesis 2–3 in the 600s.

As the Assyrians had done in the North, Nebuchadnezzar now took all survivors as prisoners, culled any with skills useful in building a new revolt (e.g., landowners, artisans, scribes, warriors, nobility, and priests), and transported them, along with their wives, concubines, and children, hundreds of miles east to the outskirts of Babylon, his royal city. There they joined their fellow Judeans, who had been deported by Nebuchadnezzar twelve years earlier after his first capture of Jerusalem in 598.

Unlike the Assyrians, who were noted for their harsh treatment of conquered peoples, Nebuchadnezzar left the new captives, along with the old ones, in relative freedom in Babylon. They were free to establish their own communities, set up small businesses, farm the land (which was irrigated and relatively fertile, unlike the land they had left), and raise children. The Israelite captives did this well, as members of both deported groups had skills uniquely suited to re-creating a coherent society from the ground up, although they lacked two elements that were significant, if not essential, to their individual and collective identities: the Temple and the land.

The Temple in Jerusalem, built by King Solomon in the tenth century, had become the center for all worship of the Israelite God, Yhwh. Only in this Temple could cultic rituals and sacrifices be carried out. Here, too, in the center of the Temple, within a stone structure named the Holy of Holies, the Israelites of Judah believed their God had come to dwell in an almost physical sense. They also believed their God's presence was especially palpable in the land. The Israelites considered that their God was sovereign and able to exercise power only within a particular land, outside of which their God was powerless.[52] Without the land, even if they continued to worship their God, the exiles doubted their God could hear and act for them.[53]

52. See the story of the leprosy-stricken Syrian general, Namaan, in 2 Kings 5:1–19.

53. *Monotheism*, the belief that there is only one god, had not yet become the dominant Israelite worldview. Monotheism was not fully adopted by the Israelites until the end of the Babylonian Exile, when the Israelites can first be recognized as the Jews they have become. From the time of Abraham and Sarah, all those who followed Abraham and his offspring adopted his god and practiced *monolatry*, the belief that though many gods existed, worship of only their one clan or tribal god was permitted. For the Israelites, that one god was the God of Abraham, Isaac, and Jacob. This movement from monolatry to monotheism, as a consequence, developed over centuries until we can see it fully articulated in the prophecies of Second Isaiah (see esp. 43:10b–11 and 44:6c) late in the period of exile (540s perhaps).

In addition, possessing their land over the centuries had become a potent symbol of the Israelites' special status with God, guaranteed by the unconditional covenant God had made with Abraham and his descendants (see Gen 12:1–3; 15:1–21; 17:1–8), a covenant that was renewed periodically right down to King David and his heirs (see 2 Sm 7:1–16). But now Nebuchadnezzar had destroyed the Temple and the city; he had captured both the land and the people. As a result of these events, the Israelites entered foreign captivity severely traumatized by war and bereft of external or internal structures to help them make sense of the chaotic fragments of their hopes, beliefs, and communal and individual identities. They were making a new home as the remnant of a conquered people amidst the power and wealth of a foreign nation whose pantheon of gods had surely overpowered their God Yhwh. How else could the Babylonian conquest be explained? In a sense, their continued identity as followers of this one God, Yhwh, was seriously imperiled as they made their way into a radically different life.

Over time, the Israelites established lives in the small towns they built in exile, recreating for themselves and their offspring a life like that they had lost in the final destruction of Jerusalem. With freedom of movement, however, and access to arable land and the markets and commerce of the city, the exiled Israelites had to interact, sometimes on a daily basis, with ordinary Babylonians. They also had to learn a new language (Aramaic, a Semitic language related to Hebrew) and were repeatedly exposed to the alien Babylonian worldview and its

Discover for yourself

To experience on some level the anger, pain, bewilderment, anguish, and sense of abandonment the Israelite deportees experienced, read Psalm 137 or Lamentations 1–2, both of which were composed fairly soon after the Exile. As you read through either piece—or both—keep in mind similar experiences of disaster and loss that have occurred among modern peoples in the last one hundred years. Choose one of these horrific events to research, and based on that research, write a dirge (a song or poem of mourning) or lamentation in the voice of the people against whom such a human-made disaster was perpetrated.

practices. By the time the first and second generation of Israelites had been born and grown to adulthood, they would have known no other home but Babylon. Their parents, elders, scribes, and priests would have grown alarmed at the familiarity and ease of interaction their offspring had developed with the Babylonian world.

The new generations would have heard the old stories of the wonders the Israelites' God had wrought, but they also would have known that Babylon's army and fabled king, Nebuchadnezzar, had demolished their parents' and grandparents' city, Temple, and land, and forced the survivors into exile—all with apparent impunity from the anger of the Israelites' God. Their lives, however, were more comfortable than that of their forebears in Judah, and they were relatively safe from marauders' invasions by virtue of their proximity to the Babylonian capital. They also were spared the regular famines and droughts that periodically afflicted the Judah of their elders, thanks to the availability of water from the Euphrates River and Babylonian skill in creating an irrigation system of canals. They would have encountered over and over again the Babylonians' foundational story, the *Enuma Elish*, as members of the deities' temple cults retold and reenacted parts of it throughout the city during religious festivals, and so the Israelite exiles would have learned this story as well as they had learned their own foundational story, Genesis 2–3.

GENESIS 1: THE PRIESTLY ACCOUNT

These younger members of the exile could easily have abandoned worship of the God of Israel and turned to worship of the gods whose power and favor seemed to have brought Babylon military and political might, wealth, and land. But most did not abandon the God of their forebears, because their priests and scribes had developed innovative religious structures and practices to rebuild a national identity strong enough to withstand constant exposure to Babylon's allure.

The priests worked on several fronts. Places where the community could periodically assemble for social events and worship were organized (today called "synagogues," from a Greek preposition and verb meaning "to gather together"). The priests and scribes also developed a simplified worship service for regular communal prayer in these synagogues to replace the animal sacrifices and complex temple liturgies of the past. The priests also reinstituted practices that had been neglected in the land of Israel, such as kosher food laws and male circumcision. These last two religious practices became crucial, daily reminders that the people of Israel alone had been chosen by their God to be holy (see Ex 19:6). Finally, the priests also introduced a day of rest and worship on the seventh day of the week, which they called "Sabbath," after the Hebrew word for "seventh." This day would function for the exiles as a time set apart from everyday activities, circumscribed by the sacred, to consciously channel all their actions toward the worship of the God of Israel. The priests thus had reimagined their identity as a people to fit their new situation and then reinscribed it in their psyches through daily and weekly cultic practices.

The priests also compiled and edited the Temple scrolls and royal chronicles that they had brought into exile. These scrolls preserved various writings, likely starting with the work first begun by the royal scribes of Kings David and Solomon in the 900s (including Gen 2–3) and continued by subsequent generations of royal and priestly scribes. These priestly exiles arranged and judiciously augmented these works with their own reflections and stories. The priestly writer, or P, intended the finished product to serve the people both as an instruction in faithfully living the covenant of Abraham and as a beginning answer to the Israelites' haunting questions: Why had this God abandoned them, allowing foreigners to ravage the land that God had long ago promised Abraham would be theirs forever? How could this God have permitted the destruction of the sacred Temple by an alien nation? And when the enemy had finished desecrating their land, Temple, and people, why had God allowed them to be dragged into bondage to a foreign land? Was their God even a god? Did their God exist?

To provide instruction and respond to their questions, P had to begin with the profound questions that lay beneath the spoken questions. Today such questions are called "metaphysical," for example: Who is God? What does it mean to be a human being? What does it mean to be a human being in relationship with God? What does it mean for me to be in relationship with other human beings? What can I expect from God? from the world? from other human beings? What does God expect from me?

These were, and still are, questions that haunt reflective human beings. Such questions cannot be easily or satisfactorily answered through didactic exposition of principles or beliefs, but they can be explored with depth and nuance in a cosmogonic story (see above). The Priestly author of Genesis 1 chose this form of writing, just as the earlier J writer had with Genesis 2–3 (see above), to offer a creation account that would challenge the worldview and human self-understanding proffered by the Babylonian cosmogonic myth, the *Enuma Elish.* The historical, social, political, and religious situations for which Genesis 2–3 had been written and then revised had changed; something different was needed to explain to the exiles the meaning and identity of their new existence. The old story was insufficient.

P had to make clear once and for all that the God of Israel reigned in the heavens, even though it may have seemed the God of their forebears had abandoned them to the Babylonian military juggernaut or, worse, been unable to prevail against their enemy's stronger gods. In addition, as the genre of cosmogony is meant to do, P shaped their creation story to remind the exiled Israelites that their existence as a people continued to reveal the meaning and purpose the Divine had given them at the very beginning.

The priestly author immediately signals in his opening verses that he is writing a cosmogony meant both to echo the Babylonian *Enuma Elish* in his hearers' ears and to challenge that epic's power to shape the Israelites' self-understanding.[54] The priestly writer keeps the liturgical pace and rhythm in Genesis 1 close to those found in the *Enuma Elish* because he intends the Israelites to hear the Genesis account during their worship in synagogue, just as they would have heard Babylonian priests proclaiming the *Enuma Elish* in the city's temples during feasts, especially during the fall new year's festival. He also uses the structure of the *Enuma Elish* as the scaffolding for the story he is shaping, hoping to undermine whatever inroads Babylonian beliefs, values, and customs have made among the exiled Israelites.

By rewriting and reinterpreting the Babylonian foundational story, the priestly writer is reinscribing the shared identity of the Israelites on their collective psyche and renewing their unique theology. P's story subverts the *Enuma Elish* in five crucial ways: (1) only God creates; (2) God creates through speech, not through sexual activity or

54. The entire *Enuma Elish* is available in English translation at several websites; the following is but one example: *http://www.crivoice.org/enumaelish.html.*

violence as in the *Enuma Elish*; (3) the world God creates is orderly, planned, and reliable, unlike the chaos that pervades the world of the *Enuma Elish*; (4) fertility of the land, creatures, and humanity comes only through God's blessing, not through the sexual activity of gods and goddesses; and (5) human beings are created on purpose and in the image and likeness of God, not as an afterthought to serve the needs of the gods. Each of these five principles comes again and again to the foreground as the priestly writer deconstructs Babylonian assumptions about the meaning of being human, the purpose of human life, and what human beings can expect to receive from God.

Analysis of the Text

In Genesis 1:3–5, P initiates three verbal patterns that he will repeat throughout the six days of creation: (1) "And God said, 'Let there be . . . ' and there was . . . ," (2) "And God saw that [it] was good," and (3) "And there was evening and there was morning, the (first, second, etc.) day." The patterned repetitions create a sense that there is order in the world and that the order is a result of God's slow and deliberate actions. The world of darkness and ocean waters (vv. 2a–b) in Mesopotamian mythology contained unimaginable terrors for humanity. But in Genesis 1, the wind or breath of God hovers over chaotic waters of the abyss and establishes that Israel's God controls what even the Babylonian gods avoid. With no more effort than it takes to speak a word, God brings light into existence and pronounces it good (vv. 3–4a). The Divine separates light from what frightens humanity—darkness—and names it "Day, and the darkness . . . Night" (v. 5). By having God name both elements, light and darkness, P asserts again, as he did subtly in v. 2, that the God of Israel controls chaos and the forces of terror. The priestly author begins to lead his Israelite audience to the inexorable and inevitable conclusion made in 2:4a, that the God of Israel outpowers, outperforms, and will outlast all generations of Babylonian gods.

In addition to the threefold verbal pattern noted above, the priestly writer employs another internal structure that links the content of what is created on the first three days to what is created on each of the next three.[55]

55. Cf. Wenham, *Genesis 1–15*, 7.

Day 1 Light	**Day 4** Lights found in the day and evening skies
Day 2 Sky separating waters	**Day 5** Birds and fish
Day 3 Land and plants	**Day 6** Animals, things that creep, and human beings
Day 7 Sabbath rest	

The initial light created and separated from darkness (vv. 3–4) on day 1 forms the unnamed sun, moon, and stars of day 4 (vv. 14–17). The birds and fish summoned into life on day 5 (vv. 20–22) fill the skies and waters first conceived on day 2 (vv. 6–8). The creation of land and plants on day 3 (vv. 9–12) anticipates the need for food, habitat, and shelter by the land-based creatures formed on day 6 (vv. 25–26). The carefully balanced symmetry of the priestly creation account signals that God intentionally plotted the arc of creation to end with animals' and humanity's creation and God's unspoken assessment of all of Creation: "God saw everything that he had made, and indeed, it was very good" (v. 31). This same forethought and symmetry are also intended by P to undermine the Israelites' attraction to a Babylonian cosmogony in which the behavior of the gods mirrors the worst of human behavior (e.g., the matricide committed by the hero of the *Enuma Elish*, Marduk).

P announces that the work of creation has been completed on the sixth day (2:1). He recalls the beginning of God's creative labors in 1:1 by proclaiming "the heavens and the earth were finished" and adds the phrase "and all their multitude" (2:1) to indicate that everything between these verbal bookends, expressed or not, finds its origin in the Divine's creative word.[56]

In the *Enuma Elish*, the haphazard, violent, and self-serving behavior of the gods and goddesses begets a world that is a by-product of deicide and conceives humanity as an indulgent afterthought. In contrast, P has Israel's God deliberately conceive and bring forth the physical world in an orderly, sequential manner. After each burst of creative activity, the writer has God recognize the inherent goodness of the physical world. Finally, P notes the creation of "a human

56. Westermann, *Genesis 1–11*, 169.

being," *'adam*,[57] who bears the image and likeness of God in both male and female versions (vv. 26–27), and who like God is standing in relationship to the rest of creation.[58]

The seventh day is not separate from the previous six days in which God performs eight creative actions (two in v. 3 and in v. 6). Rather, day 7 is the culmination of six successive days of the Divine's labor. The labor of six days of creation is only completed when P declares that God rests on this last day. With this last movement, the writer intimates to his captive audience that the seven days form one unit of time while at the same time suggesting that the seventh day has been set apart from the other six days by having been "blessed" and "hallowed" by God. This distinction between the seventh day and the other six is emphasized by the priestly writer to draw attention to the Sabbath's place within creation.[59]

P offers the captive Israelites a stark choice between the world-views of the two related stories. Which world do they want to live in? Which world do they want their children and grandchildren to live in? With which gods do they want to entrust their future? Do they want to rely on violent, unpredictable Babylonian gods to sustain them in a turbulent and barbarous world? Or do they prefer the orderly, reliable, nonviolent world created and inhabited by the God of their ancestors, who sees all of creation as "very good"?

Over their remaining years in exile, the repeated, stately recitals of this narrative by the Israelite priests, aided by newly established, or

---

57. The Hebrew word *'adam* is best rendered in English by the words "humanity" or "human being." In v. 27 P employs the Hebrew word *zacar* to indicate the individual male and *neqebah* for the individual woman.

58. God's use of "Let us" in v. 26 has often provoked comment. Why does God say "us" in reference to the Divine? This may be a vestige from an earlier pagan cosmogonic source that P never expunged, but that seems unlikely given the care P took not to retain any pagan references that would conflict with his own creation account. It may express the Israelite concept that God was surrounded by a court of lesser celestial beings, not unlike the advisors and courtiers that surrounded every serious Mesopotamian potentate. In addition, it may reflect the Israelite notion that their God is the God of Gods, and so all gods are subsumed in Israel's God, hence the first person plural. Or it may simply be an expression of someone saying to herself, "Okay, now, let's go," and moving along to another task. Cf. Westermann, *Genesis 1–11*, 144–45.

59. Ibid., 171–72.

reestablished, institutions and ritual practices (synagogue, Sabbath, male circumcision, and food laws), would undermine the Babylonian threat to Israel's identity and religious worship. The danger that the Babylonian pantheon had once exerted on the exiles' religious imagination and affections would be diminished, and the majority of Israelites, now legitimately called Jews, would hold fast to their national and religious loyalties. They and their religious descendants preserved this creation account, as well as the account found in Genesis 2–3, because these accounts continued to reveal to them truths about the nature of their God, the nature of being human, and what they might expect for themselves in a world created and inhabited by the God of J and P.

DISCUSSION QUESTIONS

1. Consider the picture of Yhwh God as presented in Genesis 2–3. How would you describe this "person"? What qualities characterize this image of God? How would you characterize the relationship between Yhwh God and the man and the woman in Genesis 2–3? What was going on in the historical, political, and social worlds of the J author that might have helped to shape the picture of Yhwh God you described above, and the relationship between Yhwh God and the man and the woman?
2. Recall the image of God the P author drew in Genesis 1. How would you describe God? What qualities characterize P's image of God? How would you characterize the relationship between P's God and the human couple in Genesis 1? What was going on in the historical, political, and social worlds of the P author that might have helped to shape the picture of God in Genesis 1, and the relationship between God and the human couple?
3. Reread the last sentence of the chapter. What truths about the nature of God, the nature of being human, and what humanity might expect for itself in a world created and inhabited by the God of J and P might you infer after having read this chapter?

GLOSSARY

canon. From the Aramaic word *qaneh*, meaning "reed." Originally, the *qaneh* was used as a yardstick, to measure objects. The term was later used when determining whether individual books "measured up" or qualified for inclusion in the Bible. Eventually the term *qaneh* or *canon* came to mean the "container" of the accepted books of the Bible. *Canon* may also be defined as the official list of the books in the Bible, or the Bible's table of contents.

canonization. The centuries-long process by which the Israelites, early Jews, and early Christians sifted first through the oral accounts to determine which to write down, then through the written accounts to determine which to make into books, and finally through these books to decide which to include and how to order those books that we now call the Bible.

cosmogony. A theory or account of the creation of the universe.

criticism. A detailed analysis of a text that seeks to make a literary or historical assessment of that text. It does not "criticize" or call into question the text's historicity or value.

Enuma Elish. The Babylonian creation myth, which dates from at least the eighteenth century BCE. Tells about the creation of the world and how younger gods under the leadership of Marduk overcame older gods under the leadership of Tiamat. *Enuma elish* are the myth's opening words.

exegesis. A technical word meaning to "draw out" the sense of a text.

fundamentalism, biblical. An approach to biblical interpretation that asserts the Bible is without error; every word must be taken in its "natural sense." Such an approach dismisses historical and literary approaches to interpreting the Bible.

historical-critical method. A methodology used to interpret the Bible that came into wide use in the nineteenth century. This method recognizes the Bible as not only inspired by God but also as a collection of ancient documents composed by numerous human beings over millennia. This method applies historical, literary, and philological analysis to the biblical text to establish what it originally meant in order to ask what the text can mean for believers today.

inerrancy. Being without error. When the notion of inerrancy is applied to the Bible, it has two possible applications: **limited inerrancy** maintains that the Bible is without error only in those things it communicates that are necessary for a human being's salvation; the biblical text may include information nonessential to one's salvation, which the contemporary world has determined to be erroneous in matters of geography, history, science, or mathematics. **Strict inerrancy** or **plenary verbal inerrancy** claims that the errorlessness of the Bible includes every word in its modern, literal sense.

myth. From the Greek word *mythos*, or "story." When used in connection with biblical stories, it connotes stories that are created to express the deepest truths of what it means to be a human being, such as the stories found in Genesis 1 and Genesis 2–3.

northern kingdom. Originally part of the Israelite nation united under the kingships of David and Solomon (1000–922 BCE) in which ten of the twelve tribes of Israel lived. When Solomon's son Rehoboam ascended to the throne, the northern tribes broke away and formed their own kingdom with a non-Davidic king. The portion of land they occupied is often referred to as "Israel." It endured for about two hundred years before it was conquered by the Assyrian Empire in 722 BCE.

oral tradition. A process of memorization and public recitation by trained specialists by which preliterate societies preserved prodigious amounts of knowledge.

Pentateuch. See **Torah.**

Scripture. Writing considered sacred by a particular religious group, usually presumed to reveal the desire of the Divine.

southern kingdom. Portion of David and Solomon's kingdom that remained loyal to Rehoboam, heir of King Solomon, when he ascended to the throne in 922 BCE. This kingdom encompassed the area in which the two tribes of Judah and Benjamin lived and in which Jerusalem is located. It was named "Judah" in recognition of the larger of the two tribes. The southern kingdom fell to the Babylonians in 586 BCE.

Torah. Hebrew word for "law" or "instruction." Also the name given to the first five books of the Bible: Genesis, Exodus, Leviticus,

Numbers, and Deuteronomy. **Pentateuch** is a Greek word meaning "five scrolls" and is another name used for the first five books of the Bible.

RESOURCES FOR FURTHER STUDY

Clifford, Richard J. *Creation Accounts in the Ancient Near East and in the Bible.* CBQMS 26. Washington, DC: Catholic Biblical Society of America, 1994.

Clifford, Richard J., and John J. Collins. *Creation in the Biblical Traditions.* CBQMS 24. Washington, DC: Catholic Biblical Society of America, 1992.

Cotter, David W. *Genesis.* BO. Collegeville, MN: Liturgical Press, 2003.

Smith, Mark S. *The Priestly Vision of Genesis 1.* Minneapolis, MN: Fortress, 2010.

Tribble, Phyllis. *God and the Rhetoric of Sexuality.* OBT. Philadelphia: Fortress, 1978.

Wenham, Gordon J. *Genesis 1–15.* WBC. Waco, TX: Word, 1987.

Westermann, Claus. *Genesis 1–11.* CC. Minneapolis, MN: Fortress, 1994.

Witherup, Ronald. *Biblical Fundamentalism.* Collegeville, MN: Liturgical Press, 2001.

chapter 2

Scientific Knowledge and Evolutionary Biology

Ryan Taylor
Salisbury University, Salisbury, MD

To critically evaluate the so-called debate between evolution and theology, one must first understand what science is and what it is not. Simply put, science is a framework for inquiry that generates knowledge about the natural world. Science is an incredibly powerful tool that has provided us with a profound understanding of the natural world. There are, however, limits to science. In this chapter, we will explore in some detail what scientific knowledge is, what evolution is, and finally what limits are imposed on scientific knowledge.

For many, the word *science* conjures up a variety of images: white mice, lab coats, glass beakers, Bunsen burners, and the like. While science often involves these things, the field of science is incredibly diverse and includes a dizzying array of approaches. Unfortunately, most scientists remain so involved with their day-to-day work that they don't take time to advance a better public understanding of their work. As a result, many non-scientists have a weak grasp of what science really is. Sadly, much criticism of science comes from those who do not understand it.

SCIENCE AND ITS METHODOLOGY

To gain knowledge through science, one must first develop an idea (or hypothesis) about how something in the world works. This hypothesis will be derived from observation and experience. Next, this hypothesis will be tested using an experiment. Testability is truly the cornerstone of science; if a hypothesis is not testable, it is not scientific. To be testable, a hypothesis must be falsifiable (able to fail). Likewise, the experiment testing a hypothesis must be designed so that its results will either support or negate the hypothesis. For example, it is conceivable that an experiment could be designed such that no matter the outcome, the hypothesis would be supported. This would not be a scientific test. Given the twin requirements of testability and fallibility, it quickly becomes clear that science is limited to discovering information about the physical world. For example, consider the claim, "God exists." This claim cannot be scientifically tested because it cannot be proved false. As such, this claim cannot be used as a scientific hypothesis. Does this make the claim untrue? No, it simply is not a scientific claim.

Another common misperception about science is that it produces a body of "facts" about the world. The term *fact* implies a certain and unchangeable knowledge about a particular subject. By this definition, much of scientific knowledge is certainly not fact. By its nature, science must follow an **inductive logic**. **Inductive logic** uses information from a set of specific examples to explain a general phenomenon. Science is limited to inductive reasoning because it is simply impossible to test all possible outcomes in the world. Because a scientist cannot test all possible cases, he or she must draw a general conclusion about the world based on a limited set of instances. This is the problem of induction.

As an example, consider a scientist who wants to know the migration path of a particular **species** of duck. The scientist will develop a hypothesis that the species follows a particular migration route. To test this, the scientist might attach GPS receivers to thirty captured ducks before the fall migration and then track the flight pattern of those individuals to their wintering grounds. If those thirty ducks follow the hypothesized route over several years of testing, then the scientist concludes that all ducks of this particular species follow that route. Time and financial constraints dictate that the scientist cannot radio track all of the tens of thousands of individual ducks that migrated in those years. So the scientist is left to draw an inductive conclusion about all ducks of this species based on the limited sample of thirty. Although unlikely, it is possible that the scientist's conclusion is wrong. Perhaps the scientist's samples were biased by the fact that she was able to capture those particular ducks. Perhaps the stronger and faster ducks (most of the **population**) were able to evade capture by the scientist and followed a different migratory route than the slower ducks. If this were true, then the scientist would be wrong about this species' migratory route.

Any scientist worth their salt understands this problem of induction and draws only a tentative conclusion. The well-trained scientist in this example would claim that her hypothesis was supported and that it is *likely* this duck species travels the hypothesized route. What the scientist does not do is claim to have *proved* this duck species follows a particular route. Science never proves or guarantees anything with certainty and this is precisely what makes science dynamic and exciting. New research often reinforces old ideas, but it also provides new discoveries and turns old "knowledge" on its head.

The role of inductive logic in scientific reasoning is an important consideration in the discussion of evolution and theology. It is valid to ask how scientists can claim that evolution is the force driving the diversity of life on Earth if science cannot prove anything. Indeed, how can scientists make any claims at all for that matter? The answer lies in the process of science.

In any scientific field, such as biology, there are many subdisciplines, each with its own set of researchers interested in particular questions. Each generation of scientists continues to test and refine the questions of their particular field. Again and again these scientists test hypotheses using different techniques. They apply the same questions to different species. They ask fresh questions to attack a problem from different angles. Often, hypotheses are refuted and rejected from the body of scientific knowledge. Sometimes hypotheses continue to hold up to scrutiny. If a hypothesis continues to withstand the scrutiny of many experiments by many different scientists over many decades, then the hypothesis comes to be considered a theory.

Evolution is often dismissed as "just a theory." It is important to understand, however, that in science, *theory* means something very different from how the word is used in everyday speech. In everyday speech, *theory* often means "just a guess," as in, "My uncle Joe has a theory on how the New York Yankees will do this season." A scientific theory is much more than just a guess; it is a claim about the world that has withstood decades of rigorous investigation by many different scientists. This means a scientific theory is a powerful claim about the world that is backed by an enormous amount of experimental support.

Evolution was first proposed in part by British naturalist Charles Darwin (1809–1882) with his publication of *On the Origin of Species by Means of Natural Selection* in 1859. Not surprisingly, his book caused tremendous social upheaval, but it also shook up the field of biology and ushered in a new era of research. Darwin's hypothesis of evolution by natural selection has withstood 150 years of rigorous and repeated testing, elevating it to the powerful status of scientific theory.

To put scientific theory into perspective, consider gravity. No one disputes that gravity is the force that keeps Earth in orbit around the Sun and causes objects to fall to the ground. No one has ever observed gravity directly—one can only see its effects, devise experiments to test it, and model it with mathematics. Gravity therefore is not a proven fact but "only" a scientific theory. When scientists discuss a theory, such as evolution by natural selection, they are discussing a conceptual explanation of the world that is supported by a huge volume of solid evidence.

THE HISTORY OF EVOLUTIONARY THEORY

Evolutionary biology is a field of science. Knowledge regarding evolution proceeds by the scientific process of developing and testing hypotheses. One meaning of the term *evolution* is simply "change." When used in a biological context, *evolution* refers to change in living organisms that occurs over the course of many generations. It is now understood that genetic change is the underlying mechanism of evolution, an area explored in detail later in this chapter.

Although today Darwin is universally acknowledged as discovering how evolution proceeds by natural selection, many people made early and important contributions to evolutionary thought. Jean-Baptiste Lamarck (1744–1829) proposed that species change over time as a result of the use of a particular part of the body. For example, he proposed that giraffes have long necks because individuals stretched repeatedly to reach leaves high on a tree. This resulted in a giraffe's neck becoming longer over the course of its lifetime, and the long-necked trait, Lamarck reasoned, would then be passed on to its offspring. This would be akin to claiming that if a person were involved in an accident and lost a finger, then the person's children would be born also missing that finger. Lamarck's proposed mechanism for evolution was later shown to be largely false, but his ideas generated much interest in evolutionary thought.

Another notable contributor to evolutionary thought was Darwin's contemporary, Alfred Russell Wallace (1823–1913). Wallace developed the concept of natural selection independently of Darwin, although today Darwin is given the lion's share of credit for evolutionary theory because of his publications and the in-depth nature of his numerous observations and experiments.

From 1831–1836, during his voyage around the world aboard the HMS *Beagle*, Darwin began forming his concept of natural selection. He found a variety of fossils that suggested evolutionary change; many species in these fossils resembled living species but were larger or had other features that were distinct from living organisms. In the Galapagos Islands, Darwin also observed living species that looked similar to those on mainland South America but that had other, unique characteristics. Further, he noticed that there were distinct differences between similar bird species on the neighboring islands of the Galapagos. These differences among birds, some living within sight of each other, struck Darwin as odd. He reasoned that if God were responsible for creating all organisms in their present form, it would seem a wasted effort to create extremely similar yet distinct species on islands in such close proximity. Upon returning from his voyage, Darwin continued to develop his concept and compile support for his idea. After some twenty years, he finally published his groundbreaking *On the Origin of Species*. The extraordinary rigor and large body of amassed evidence in his book propelled the concept of natural selection to the forefront of scientific thought.

Shortly after the publication of *On the Origin of Species*, many biologists adopted Darwin's ideas about natural selection and common descent. What was not yet understood, however, was what led to variation among individuals or why offspring tend to look like their parents (heritability). This was an important criticism of natural selection, and while many scientists of the late 1800s adopted Darwin's ideas, there were also many critics. It was not until the 1930s that a group of scientists applied the field of genetics to the concept of natural selection. This provided a solid foundation to explain inheritance and clarified how biological evolution could occur through genetic change based on observable variations in natural populations. These discoveries in the 1920s were termed the "modern synthesis" and ushered in a more complete understanding of evolution, now sometimes referred to as the neo-Darwinian evolution. This neo-Darwinian paradigm generated new research programs and unified formerly isolated fields such as genetics, anatomy, and ecology.

In the roughly sixty years since this modern synthesis, scientists have studied evolution on many different levels and made enormous strides in understanding the process of evolution. Molecular biologists have examined how proteins evolve. Population geneticists have examined how **genes** evolve within populations. Anatomists and organismal biologists have studied phenotypic (visible traits of organisms) evolution. And geologists and paleontologists have contributed to the understanding of species change over geologic time. Thousands of scientists have filled numerous volumes with rigorous evidence for evolution, and this evidence continues to grow every year.

THE MECHANISMS OF EVOLUTION

Evolution can proceed by several different mechanisms. One of the most important of these, natural selection, was described by Darwin before the discovery of genes. Despite not knowing that genes are responsible for heritable traits (i.e., traits that are passed on from parents to offspring), Darwin recognized that (1) variation exists within populations of organisms, (2) traits are heritable, and (3) individual organisms, with their own unique traits, have different rates of reproduction. These three conditions drive natural selection.

On the face of it, the process of natural selection is remarkably simple. Nearly all populations of organisms show some level of variation. This variation can be genetic and can manifest in obvious phenotypic differences (see figure 1). Darwin recognized that natural

IMAGE: RYAN TAYLOR

Figure 1. Phenotypic variation within a species. The three shells in this image belong to the same species of conch snail, but show color and size variation. Both genetic and phenotypic variations are present in most populations of organisms.

forces act on this variation, resulting in some individuals that survive and reproduce and others that die and fail to reproduce. Among individuals that survive, some produce more offspring than others. As an example, consider a population of deer that shows variation in leg length (and hence running speed). Individuals with longer legs are more likely to outrun wolf predators than their shorter-legged counterparts. A greater number of long-legged than short-legged deer thus survive and reach reproductive maturity. As a result, there are more long-legged than short-legged deer available to reproduce in the breeding season. Each reproducing deer passes on its genes (and associated **phenotypes**) to its offspring. Since more long-legged deer survive and reproduce, there are more long-legged than short-legged offspring produced. In the next generation, therefore, the average length of the legs will be slightly longer, but there will still be some variation in leg length among individual deer. Again, as wolves hunt the deer, they will prey most often on those with the shortest legs. Thus, with each passing generation, the average leg length will become slightly longer (see figures 2a and 2b).

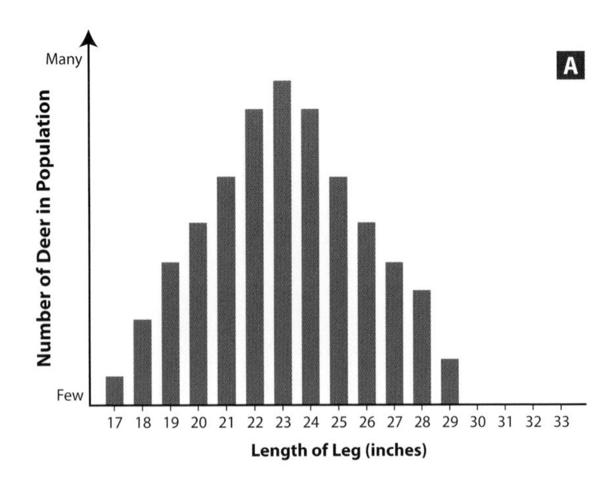

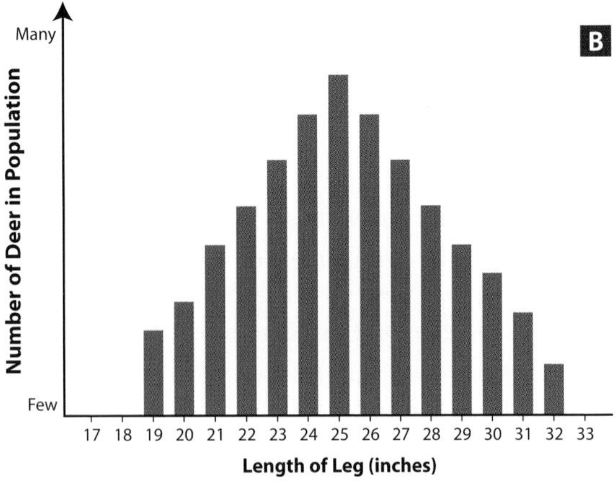

Figures 2A/B. Evolution of leg length in a deer population subject to predation pressure by wolves. (A) This graph shows the distribution of leg length that might be typical in a deer population. A small number of individuals have very short legs (17–18 inches), a small number have very long legs (28–29 inches), but most individuals have intermediate-length legs. In this population, the average is 23 inches. (B) This graph shows the distribution of leg lengths in the same deer population many generations later, if the deer with the shortest legs are eaten more frequently by wolves. Fewer short-legged individuals survive long enough to pass on the short-legged trait to their offspring. Over many generations, the average leg length tends to increase. Notice that there are now no deer in the population with 17- to 18-inch legs and a small number of deer with legs of 31–32 inches in length. The average leg length in the population has also shifted to become slightly longer at 25 inches. (Author illustration)

Darwin recognized that if this sort of gradual adaptive change continued long enough, the species would eventually change sufficiently to warrant being described as a completely different species. This sort of adaptive change is what Darwin referred to as "descent with modification." If an organism possesses a trait that improves survival and reproduction, it is considered better adapted to its environment than other members of its species. Over time, traits that are advantageous to survival become more common, and the species undergoes modification of traits with each passing generation. Biological evolution is the change in the frequency of genes contained in organisms within a particular population from one generation to the next.

Although Darwin did not know it at the time, every living organism possesses a set of genes that are responsible for passing traits from parents to offspring. The set of genes that each organism possesses varies to some degree between individuals. This is most evident when one looks at other humans. There is a tremendous variation in skin color, eye color, height, and so on. Most of this variation is due to differences in genes from one individual to the next. Individuals of a particular species (humans included) all have the same genes but often have different **alleles**. Alleles are the variants of one particular gene. Consider, for example, a gene that contributes to eye color. One allele may specify brown eyes and another, blue. Both are genes that control eye color, but each produces a different color. These different versions of the same gene are called alleles.

Genes (alleles) produce certain traits by creating proteins that perform a particular job in an organism's body. For example, there are proteins that provide structural support (like the proteins that make up your hair and fingernails). There are proteins that act as enzymes that help run an organism's metabolism. And there are proteins that help to maintain the biochemical environment needed to sustain life. Proteins are essential for nearly every aspect of the structure and function of living organisms.

The process by which genes build a body is quite complex, and even a single gene can have a profound effect on outward traits (the phenotype). Take, for example, the size of dogs. Current research suggests that just a single gene determines the size of domesticated dogs, producing the difference in size between a toy poodle and a

Great Dane, for example. Multiple genes may also work together to produce one trait. What is now understood is that different combinations of alleles produce variations of the phenotype, and these variations confer many possible advantages or disadvantages for the organisms that bear them.

Let us return for a moment to the example of wolves hunting deer. The variation in deer leg length is likely due to the variation in alleles for this trait that are present in the deer population. Suppose that eight alleles dictate the leg length in a deer. If a deer possesses a combination of eight particular alleles, it may have longer-than-average legs. If another deer has a combination of eight different alleles, it may have shorter-than-average legs. This begs an interesting question: why is there variation? There are really two answers to this question; the first is *genetic shuffling*. When two sexually reproducing organisms mate, there is a random shuffling of genes that are selected and passed on to the offspring. Since the mother and father each contribute only half of their genes to the offspring, the offspring inherits a random set of genes from each of its parents (see figure 3).

As an analogy, consider that you have two decks of cards, where each deck has eight cards numbered three through ten. Each deck represents the **genotype** of an individual deer parent and each card represents one allele that determines leg length of the deer. Each deer parent will contribute half of its genes to the offspring during mating. If the two card decks are shuffled and then four cards from each deck are dealt, thus forming a new deck of eight cards, the new deck would represent the genotype of the deer parents' offspring. Since each of the parental decks was shuffled and the cards dealt randomly, numerous combinations of cards would be possible. Consider a new deck combination (genotype of offspring) consisting of two threes, two fours, two fives, and two sixes. If the low cards (alleles) represent shorter legs, then this offspring would have shorter-than-average legs. If we have a combination of two sevens, two eights, two nines, and two tens, then this deer would have longer-than-average legs (alleles represented by higher value cards). Of course, both of these extreme combinations are unlikely. What is more likely to occur in the offspring is a random combination of low, medium, and high cards, resulting in a leg length somewhere around the average.

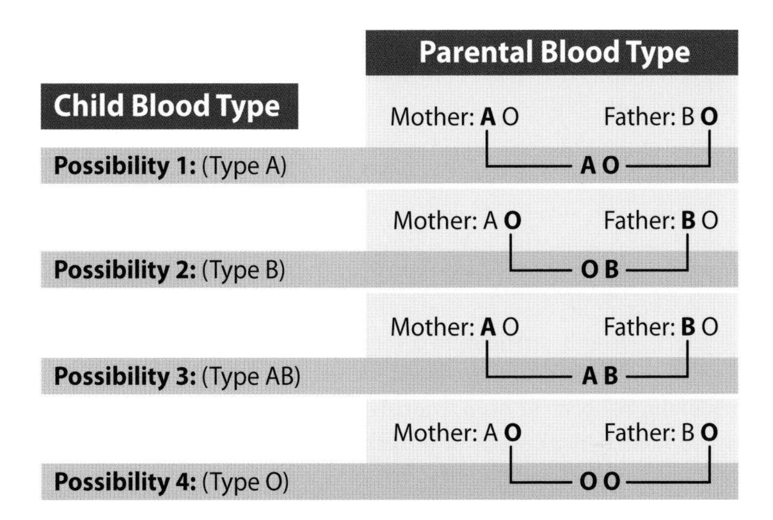

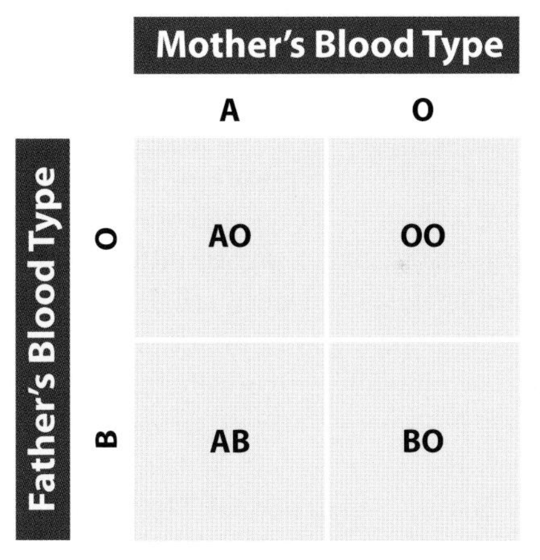

Figure 3. An example of how sexual reproduction generates random shuffling that results in genetic variation. Three alleles affect blood type in humans, producing blood-types A, B, and O. Because every person contains two blood-type alleles (one from the mother and one from the father), several combinations can arise. Consider two parents, one with type A and the other with type B blood. The allele producing blood-type O is recessive; therefore, if a person has A or B as their other allele, they will have type A or B blood. Each parent can donate only one allele to the child; this means the child could have one of four possible blood types. The first diagram depicts the contribution of alleles from parent to child; the second diagram shows the same information written as a Punnett square (the standard format for working out possible gene combinations of offspring produced by parents from sexual reproduction). (Author illustration)

If we shuffled many pairs of such decks and dealt the cards to represent all the deer born into the population in a particular year, the deer offspring would end up with leg lengths that follow a roughly normal distribution (see figure 2a). A small number of individuals would have very short legs, a small number would have very long legs, and the majority would fall somewhere in between. Those individuals with longer legs would have an advantage in that they are more likely to evade wolves and survive to reproduce. Using the card example, consider that individuals in the bottom 10 percent of the leg-length distribution are eaten by wolves. These individuals had proportionally more lower-value cards. If one then takes the remaining 90 percent of individuals, shuffles their decks, pairs them up, and deals out new sets of offspring, the new generation now will have, on average, a slightly higher card value (because proportionally more lower-value cards have been removed from the deck—the representative gene pool). Thus, this new generation of deer will have slightly longer legs than the previous generation.

The second cause of variation in populations is **mutation**. When cells within organisms divide, the dividing cells must make new copies of the genes (a process called **replication**) that go into the new cell. Sometimes during replication a copying error occurs (i.e., a mutation) and the replicated gene receives a slightly different code. The mutated gene copy goes into the newly divided cell. If that mutated gene goes into a sperm or an egg cell, it may be passed on to the offspring during reproduction. These mutations are usually harmful. In extreme cases, the offspring with the mutation may die *in utero*. In less extreme cases, the offspring with the mutation will be born handicapped in a way that causes it to die earlier than normal. This early death may be due to any number of problems, including being unable to feed well or an increased susceptibility to disease. An early death means the individual with the mutated genes will either produce fewer than average offspring or none at all. When this happens, the mutation disappears from the population rather quickly, usually within a few generations. In rare cases, however, the mutation may be beneficial. If the mutation is beneficial because it confers some survival or reproductive benefit, then the individual will tend to leave more than the average number of offspring. In this situation, the mutated gene spreads and becomes more common in the population with each generation.

Mutation is the ultimate source of all genetic variation. The mutation of genes generally occurs at random and is not correlated with known factors affecting wild populations (chemicals that cause mutations resulting in cancer are an exception to this). One common misunderstanding about evolution involves the nature of the random process of mutation. It is commonly assumed that the entire process of evolution is random. Mutation is random, but natural selection is not. Within a population, genes occasionally mutate and are then thrust into the world to be "tested." If the gene provides a survival and hence reproductive advantage, the gene and its associated trait will become more common over time. If the gene is not beneficial for survival and reproduction, it is quickly eliminated from the population.

To summarize, mutation and genetic shuffling in sexual reproduction create random genetic variation. But natural forces such as predation, disease, and weather act on this variation and favor those organisms that are best suited to survive and reproduce within a particular environment. Individuals carrying beneficial genes (those best adapted) tend to produce more offspring and over many generations, the beneficial genes become more common in the population. If a gene is harmful, organisms bearing it produce few or no offspring, so over time, the gene disappears from the population. Thus, as this process is repeated over thousands to millions of generations, the environment affects which genes an organism is likely to carry and hence determines its appearance and physiological function. Given enough time, the small, adaptive (genetic) changes that are produced by environmental selection accumulate and can lead to dramatic changes in living form and function.

There is no set amount of time required for evolutionary change to take place. The amount of time required for change depends on the type of change and the generation time. For example, a genetic mutation can crop up in just two generations if a copying error is made during DNA replication and then passed on to an offspring. An important concept to keep in mind is that individuals do not evolve; change can only occur across generations. Thus, the faster an organism's generation time, the faster the species can evolve. A species of bacteria that can reproduce every twenty minutes will exhibit faster rates of evolution than an elephant species with a generation time of twenty years or more. Changes to large sections of an organism's genome or the evolution of new species often take many thousands to millions of generations. Biologists consider changes that take place over tens of thousands of years to be rapid; changes occurring within species lineages that occur over millions of years are typical.

So far we have considered evolution as it occurs within one species over time. But how might evolution create a new species? Many closely related living species share similar traits: take snail species that build seashells, for example. Closely related snail species share features in common but also have unique characteristics that distinguish them from other snail species (see figures 4a and 4b). One of the most commonly accepted definitions of *species* is "individuals that can interbreed to produce viable offspring." If two organisms cannot produce viable offspring, they are considered to be of different species.

Several mechanisms can account for the evolution of many closely related species, which is known as **adaptive radiation**. One such mechanism is called **allopatric speciation**. For allopatric speciation to occur, a barrier must form within the geographic range of a particular species. This barrier could be a mountain range that is rising from geologic uplift, a river changing course, a glacier moving down a continent, or a variety of other natural processes. The barrier splits the species into two separate populations. If the barrier is sufficient to prevent migration between populations, then the populations do not interbreed. Over time, each population adapts to the conditions of its local environment. If at a later time the barrier is removed (the mountain range erodes, the river shifts course, etc.), then when the two populations again intermingle, they may have evolved genotypes that are sufficiently different to prevent interbreeding.

The formation of a mountain range provides a good example of how allopatric speciation works. When a mountain range pushes up near the coast, it often creates different habitats on either side. The coastal side is typically wet with frequent rainfall forming from ocean evaporation. The mountains block some air and cloud flow, creating a much drier habitat on the opposite side. Higher elevations are typically much colder than lower ones. Such a temperature difference would be sufficient to prevent many plant and animal species from crossing the mountains, and the populations split by the mountain range would adapt to the different habitats created by the mountain

Figure 4a. The shells in this image represent the variation between distinct, but closely related, cone snails. They share common characteristics (note the similar shape and pattern of the shells), indicating that they evolved from a common ancestor. They do not interbreed, however, and thus represent distinct species.

Figure 4b. These shells illustrate the variation between distinct, but closely related, auger snails. They share common characteristics (note the similar shape and pattern of the shells), indicating that they evolved from a common ancestor. They do not interbreed, however, and thus represent distinct species.

range. Another excellent example of allopatric speciation was created by the closing of the Isthmus of Panama. Closing the isthmus created different habitats in the formerly homogenous ocean. As Panama rose from the sea, the ocean on the Caribbean side developed into a clear, tropical sea, while the ocean on the Pacific side developed into a cooler, more turbid environment. Today, there are many "sister" species on each side of the isthmus that look similar but have evolved traits more suited for the marine environment on that side of the isthmus.

Natural selection is one of the most important mechanisms driving evolution. Other processes, however, can also drive evolutionary change. One of these is genetic drift. Genetic drift is a random process in which a particular allele by chance fails to be passed on to the next generation. This is most common when alleles are somewhat rare and the population is relatively small. Consider, for example, a population of deer in which there are only 100 individuals. Of these individuals, only two have a copy of a particular allele, let's call it allele X. In sexually reproducing populations, not every individual mates every year. If the two individuals with allele X do not reproduce and then die before getting the opportunity to reproduce in the following year, then allele X would be permanently lost from the population. Genetic drift generally does not drive adaptive changes that result in the appearance of new traits. But it can be an important factor that results in the loss of a trait or a process that reduces genetic variation in a population.

Another process driving evolution is sexual selection. Scientists have paid tremendous attention in recent decades to this process by which showy or conspicuous traits can evolve. A familiar example is the oversized tail of peacocks. Interestingly, the peacock's tail not only makes it a more visible target for predators but may also limit the male's ability to escape as it is burdened by a heavy train of feathers. Peahens are not troubled with such tail feathers. On the face of it, these showy traits seem to present a problem for the theory of natural selection. If natural forces continually shape organisms to be better suited for survival, then how can traits that reduce survival evolve? Darwin was extremely concerned about this and the problem that it presented to his theory of natural selection. He spent much time working on this problem and in 1871, proposed the mechanisms of sexual selection in his book, *The Descent of Man and*

Selection in Relation to Sex. Darwin described two processes by which large or showy (and seemingly maladaptive) traits could evolve.

The first of these mechanisms of sexual selection is what biologists typically refer to as female choice. Under this system, females do not randomly mate with males, but instead carefully select the males with whom they mate. This makes sense from an evolutionary standpoint. Females invest more in each act of reproduction. Eggs require more energy to produce than sperm. Further, in the case of mammals, females must gestate and feed their offspring for a considerable time. If a female makes a poor mate choice—selecting a male who abandons her to care for their offspring alone, for example—she has lost much more than would a male who makes a poor mate choice, such as mating with a genetically inferior female. In most cases, the male can abandon her to mate again. Because of the added time and energy constraint, females of most species simply cannot pass on their genes via reproduction as quickly as males. This sets up a situation in which females are selective about their mate choice and thus force males to compete for them. This nonrandom mating occurs when females choose to mate with males that have the showiest or most conspicuous courtship displays. A classic example is the research finding that peahens prefer to mate with peacocks whose tail fans are larger and have more eyespots.[1] Males with smaller-than-average tails are less likely to mate and will leave fewer offspring in the next generation; conversely, males with the largest tails will win more matings and produce relatively more offspring. Male offspring will be more likely to possess their father's larger tails. Over many generations, this female preference will drive the evolution of larger and larger tails in males. The peafowl are just one example of this process of female choice. This same process has been demonstrated in African widowbirds,[2] swordtail fish,[3] guppies,[4]

1. M. Petrie, T. Halliday, and C. Sanders, "Peahens Prefer Peacocks with Elaborate Trains," *Animal Behavior* 41 (1991): 323–31.

2. M. Andersson, "Female Choice Selects for Extreme Tail Length in a Widowbird," *Nature* 299 (1982): 818–20.

3. A. Basolo, "Female Preference Predates the Evolution of the Sword in Swordtail Fish," *Science* 250 (1990): 808–10.

4. A. Kodric-Brown and J. H. Brown, "Truth in Advertising: The Kinds of Traits Favoured by Sexual Selection," *American Naturalist* 124 (1984): 309–23.

spiders,[5] frogs,[6] and other species. In virtually every animal species where males possess some sort of increased color, ornamentation, or courtship display that females lack, there is evidence that female mating preferences have driven the evolution of the male trait.[7]

A second process of sexual selection occurs through male–male competition. In this process, males compete directly for females by fighting; thus, traits that provide males with a fighting advantage tend to be favored. A good example of this is antler size in deer. In most species, only males produce antlers and these are weapons for fighting. There is often considerable variation in antler size among males. Larger antlers provide better leverage for the shoving matches that males engage in; males with larger antlers tend to win fights. Those males that win these contests gain access to many females (the harem) and often sire multiple offspring. The losers often do not mate at all. Thus, large-antlered males tend to leave most of the offspring, and their offspring of course will have relatively large antlers.

Although large or showy ornamentation is likely to be detrimental to survival (consider a peacock trying to escape a fox while dragging a long tail train), these ornaments also confer a reproductive advantage. Evolutionary advantages are often spoken of in terms of survival and reproduction. Survival is obviously important; organisms cannot reproduce unless they are alive. In reality, it is only reproduction that matters. Consider the example of a population of peafowl. If males with the largest tails live an average of two years before being eaten by foxes, but sire forty offspring per year, then they leave eighty offspring in their lifetime. If males with the smallest tails live an average of four years, but produce only ten offspring per year, then they typically leave forty offspring in their lifetime. Even though the long-tailed males live significantly shorter lives, they are leaving proportionally more long-tailed offspring. Thus,

---

5. E. A. Hebets and G. W. Uetz, "Leg Ornamentation and the Efficacy of Courtship Display in Four Species of Wolf Spider (*Araneae: Lycosidae*)," *Behavioral Ecology and Sociobiology* 47 (2000): 280–86.

6. H. C. Gerhardt and F. Huber, *Acoustic Communication in Insects and Anurans: Common Problems and Diverse Solutions* (Chicago: University of Chicago Press, 2002); M. J. Ryan, *The Túngara Frog: A Study in Sexual Selection and Communication* (Chicago: University of Chicago Press, 1985).

7. M. Andersson, *Sexual Selection* (Princeton, NJ: Princeton University Press, 1994).

the average tail size of males in the population increases over time because proportionally more long-tailed genes are produced in the population with each generation.

Based on these processes of sexual selection, it would seem that male traits would always get larger and showier with each passing generation. Multiple evolutionary processes are often at work, however. In many cases, an upper limit may be placed on male ornamentation by the laws of physics dictating that at some point a male may simply become unable to carry the ornaments around. This is what is thought to have happened to the Irish elk. This species went extinct because their antlers grew to such an enormous size that they placed too great a burden on the animal, contributing to the demise of the species. Alternatively, predators may place an upper limit on the evolution of a trait. In many cases a balancing selection is reached between the trait increasing male mating success and the trait increasing the probability the male will be eaten by a predator. As an example, the male túngara frog of Central and South America produces either a simple or a complex courtship vocalization. The complex vocalization is more attractive to females and increases the male's chance of mating,[8] but it also attracts bat predators and increases the male's chance of being eaten. While the complex vocalization evolved through female mating preferences, its continued evolution may be limited by predators.

Another factor that may limit male ornamentation and contribute to variation is that alternative mating strategies may be equally successful. In some cases, being bigger and more colorful is not always the best strategy. In sunfish, for example, some males grow to be larger and more colorful than other males. These males are the preferred mates of females, and these larger males attract the vast majority of females in the population. But other males retain a small size and drab coloration upon maturity. These males have adopted a sneaking strategy that is quite effective. When a large male is courting a female, the small male slides in between the male and the female—in essence behaving like another female. The large male is fooled and continues to court the two fish as if they were both females. When the real female deposits her eggs in the nest,

---

8. Ryan, *The Túngara Frog*.

the small sneaker male quickly dumps his sperm, fertilizes the eggs, and leaves the large male to care for his offspring.[9] In short, multiple evolutionary processes often work in tandem to limit the evolution of male sexually selected traits (e.g., by predation) or, in some cases, increase the diversity of male traits (e.g., multiple mating strategies).

EVIDENCE FOR EVOLUTION

With the exception of some of the experimental studies cited in the section on sexual selection, this chapter has provided mostly conceptual explanations of how evolution works. The first section of the chapter also noted that a scientific theory is supported by a tremendous volume of evidence. The remaining portion of this section will consider the evidence scientists have for evolution.

One important piece of evidence suggesting that organisms undergo evolutionary change is the fossil record. The fossil record not only tells us that species which once lived became extinct (e.g., the dinosaurs) but also supplies a remarkably complete record of the evolutionary change that many organisms have undergone in the history of our planet. For example, there are a series of transitional fossils that show the evolution of reptiles (dinosaurs) into birds. Examples of these include dinosaurs that had forelimbs that looked like the forelimbs of reptiles but also had rudimentary feathers. The anatomical study of living birds and reptiles shows that scales and feathers develop from the same tissues, indicating that reptilian scales were modified over time into feathers. It is thought that the evolution of feathers from scales probably provided a thermal advantage for reptiles in a cooling climate. The selection pressures that caused feathers to evolve from scales are somewhat speculative, but the fossil record clearly shows the small steps whereby scales were modified into ever larger feathers. Beyond feathers, the fossil record shows a variety of other intermediate steps in the reptile-bird lineage. For example, modern birds do not have teeth as reptiles do. But the fossil record shows several species of early birds that had teeth; these prehistoric animals also had poorly developed wings that show

9. M. R. Gross, "Sneakers, Satellites, and Parentals: Polymorphic Mating Strategies in North American Sunfishes," *Z. Tierpsychol* 60 (1982): 1–26.

intermediate characteristics between the forelimbs of reptiles and the wings of modern birds. Another example of a recent fossil discovery is that of a turtle that possessed only the bottom half of a shell. This find suggests the evolution of the two-piece shell of modern turtles developed slowly over time.

In addition to providing a time line for the evolution of many plant and animal species, the fossil record is replete with examples of intermediate human species. There is a clear evolutionary lineage from early hominid ancestors to modern humans. In this record, there is clear evidence that human brain size has increased over the last several million years. Facial bones and teeth have reduced in size. The vertebral column has adapted to bear weight while standing on two feet instead of four. The feet have evolved arches that aid in distributing weight, another important adaptation for standing on two feet.

When examining fossils and trying to reconstruct the evolutionary history of a particular group of organisms, it is critical to accurately date the time of fossilization (i.e., when the organism died and became covered with sediment). Scientists have a variety of tools to provide accurate time estimates of fossilization. One such tool is carbon dating. Earth's atmosphere contains different isotopes (forms) of carbon atoms. Living organisms accumulate these isotopes in their bodies in proportion to the ratios found in the atmosphere. This occurs because plants take up carbon in the form of carbon dioxide. Animals that eat the plants then take up carbon from the plants and thus the carbon works its way up the food web. When an organism dies, one form of carbon isotope begins to undergo decay; that is, it chemically converts to another form of carbon at an extremely steady rate. By measuring the ratio of these two carbon isotopes, scientists can get a fairly accurate date of when a fossil was formed. Carbon dating is accurate for fossils up to about 50,000 years old. For older fossils, scientists must use different chemical isotopes that decay at slower rates. Once fossils are dated, the physical features of the fossils can be correlated with the time in which the organism lived. In this way, an accurate time line of evolutionary change can be made.

Comparative biology is another powerful tool for understanding evolutionary relationships between organisms. With anatomical features, it is easy to see how bones, for example, have been modified over evolutionary time. Scientists can compare skeletal features of living organisms to each other, as well as make comparisons to extinct animals from the fossil record. For example, the same set of forelimb bones is present in all living vertebrates. While the shape of these bones has been modified for particular functions (wings for flying versus forelimbs for walking), the same bones are present in the same location in all vertebrates. There are three primary bones in the vertebrate limb: the humerus in the upper arm, and two forearm bones, the ulna and radius. The evolution of these bones is also apparent in extinct animals, and the development and modification of these bones are seen in the intermediate fossils of fish that evolved stout fins for crawling out of water. Early amphibians evolved from fish and show even more developed forelimb bones; the restructuring of the same bones is charted throughout the fossil record of living vertebrates. Even plants show clear changes in structure between ancient plants and those that evolved more recently. For example, primitive plants lacked a vascular system and seeds, which are both seen in more modern plants. Intermediate examples of plants have a vascular system, but no seeds. Then there are plant species that have a vascular system and seeds, but no flowers. Finally, modern flowering plants have all of the previous structures plus flowers.

One of the most powerful pieces of evidence for evolutionary change has come in more recent years with the advent of molecular technologies. Scientists now have the ability to determine the actual sequence of DNA code and thus can compare evolutionary changes in the genes of organisms. Not surprisingly, the more closely related organisms are, the more genes they share. For example, approximately 99 percent of the genes in humans and chimpanzees are identical. Humans share slightly fewer genes with monkeys, fewer still with dogs, and fewer still with fish.

Previously in this chapter, it was noted that evolution is genetic change over time. Small genetic changes (**microevolution**) may not lead to obvious differences in phenotype, but over time these accumulated changes can lead to dramatic changes in the phenotype and to speciation. Scientists can now measure microevolutionary changes. Within the field of population genetics, tissue samples are taken from organisms over multiple generations and in different geographic locations to compare the genetic differences that occur geographically and over time.

IMAGE: RYAN TAYLOR

Figure 5. The snail shells (volute snails) in this image represent a species complex, whereby each specimen represents a snail that possesses some unique traits that distinguish it from closely related individuals. The individuals show some level of interbreeding, however, producing intermediate forms that indicate they are not wholly unique species.

Another powerful piece of evidence for natural selection is the existence of species complexes (see figure 5). These are groups of species so similar it is difficult to determine where one species stops and another begins. Biologists who study such groups often argue among themselves about whether or not two forms of the organism represent two distinct species. In many of these species complexes, organisms share many of the same traits yet also have distinct characteristics. They may show distinct genetic differences as well. In many of these complexes, however, the representative organisms often freely interbreed. This produces many intermediate forms between the already-similar species. The difficulty in determining species boundaries points to the gradual nature of evolutionary change. It indicates that members of species complexes are currently undergoing evolution but have not yet evolved into easily defined species.

The relatively recent appearance of antibiotic-resistant bacteria is an example of evolutionary change that occurred within a couple of human generations. As antibiotics were frequently prescribed over the last several decades, many bacterial species evolved that are resistant to antibiotics. The genetic variation in bacteria populations included some strains that were not killed by antibiotics. Strains of bacteria that lived by infecting humans but were killed by antibiotics have largely been eliminated by antibiotic use. Surviving strains, which were once less common in bacterial populations, have benefitted from the elimination of the susceptible strains. These resistant strains are able to survive to some degree in the human body in the presence of antibiotics and no longer have to compete with the other strains for access to the cells they infect. Bacterial genes that provide resistance to antibiotics impart a survival benefit; hence, those genes are now more common in bacterial populations. Antibiotics provided the selective agent for the evolution of bacterial strains that cause infections that are increasingly difficult to treat.

Evidence discussed in this section represents only a tiny fraction of the empirical evidence that supports evolutionary theory; indeed, such evidence fills many volumes. It is this evidence that scientists must rely on as a guide to understanding the natural world. As a stand-alone theory, evolution provides a conceptual framework that explains the diversity of life on Earth. More importantly, evolutionary theory weaves together all subdisciplines of biology, providing a unified understanding of life.

LIMITS OF SCIENTIFIC KNOWLEDGE

Science and the field of evolutionary biology provide a profound understanding of the world. There are limits to scientific knowledge, however. Scientific knowledge requires empirical evidence. This specifically limits scientific knowledge to the understanding of processes occurring in the physical world. Science cannot make claims about the nature or existence of God. Science is mute on the existence of the soul. Science also has little to say about the subjective nature of a person's conscious experience. If one understands this, it is clear that science cannot be used as grounds for dismissing the existence of God. Some atheists claim that "evolutionary biology explains the process of life and therefore God does not exist." This argument is false because it steps beyond the limits of science. Yes, evolutionary theory provides a unifying explanation of life on Earth; however, empirical scientific evidence cannot be used to support claims about things that are not physical in nature. Conversely, the argument that

"I believe God created the world; therefore evolution is false," is equally wrong. This argument simply says faith trumps all evidence, and sadly enough, closes the door to knowledge of the extraordinary processes at work in our natural world.

This chapter has provided an overview of basic evolutionary principles and of the scientific foundation on which these principles rest. The next chapter will consider how theology and evolutionary biology can work together to form a context far richer than the contentious worldview that accepts only evolution or faith.

COMMON ARGUMENTS AGAINST EVOLUTION FROM CREATIONISM AND INTELLIGENT DESIGN

Creationism is the literal interpretation of the book of Genesis; holders of this view claim that God created the world in its present form. As such, all living organisms on Earth have remained unchanged. The theory of evolution has been strenuously criticized partly because of the fear that it denies the existence of God and therefore is a destroyer of faith. Acting on this fear, ardent believers in creationism have voiced a variety of criticisms that claim to discredit evolution. All of these so-called scientific criticisms are either borne of a misunderstanding of evolutionary processes or are a blatant distortion of science. All of the examples that have been promoted by creationists as proof of "holes" in evolutionary theory have been soundly rejected by science.

Several recent books provide point-by-point explanations of the evidence for evolution and why creationist arguments against evolution are unsound. The primary goal of this chapter is to provide a solid basis for understanding science and evolution. As such, this chapter has focused primarily on scientific theory and has not strayed into discussions of the arguments against evolution. For readers who may still question the scientific validity of evolution, the following is a list of popular creationist arguments and the reasons why these arguments fail. (There are many more objections to evolutionary theory, all of which also fail to refute the theory.) To explore this topic in more detail, the text by Scott (2004) cited on page 72 under "Resources for Further Study" is recommended.

Popular Creationist Arguments against Evolution

1. **Evolution violates the second law of thermodynamics.**

   The second law of thermodynamics states that matter in the universe tends toward entropy (a state of disorder). Living things are obviously well organized; hence, this objection contends that evolution violates this law of physics. What this objection ignores, however, is that matter in the universe tends toward entropy only in the absence of an input of energy. The Sun provides Earth with enormous amounts of energy each day; thus, the organization of life does not violate the second law of thermodynamics.

2. **Biologists do not even agree on evolution.**

   Biologists are nearly unanimous in agreeing that evolution is the process that has brought about the diversity of life on Earth. It is often pointed out that biologists argue among themselves about evolution. This is true. They are not, however, debating whether evolution has occurred. What they usually argue about are such things as the mechanisms of evolution. They argue over whether data support one mode of evolution or another. Biologists also dispute what constitutes a species and what ultimately separates one species from another. Such disagreement is actually evidence for evolution because gradual change makes it difficult to define with certainty where one species stops and another begins. Biologists almost never disagree on the theory of evolution, however.

3. **There are no transitional fossils in the fossil record showing the changes that lead from one species to another.**

   This argument is woefully outdated. Decades ago, there were tremendous gaps in the fossil record, suggesting a lack of evolutionary transition between organisms. Over the last several decades, however, paleontologists have discovered an extraordinary number of transitional fossils in many lineages. For example, there is now a complete series of fossils detailing the evolutionary history of whale species; the numerous whale species alive today all evolved from one species of land-dwelling mammal that evolved into an aquatic form that then evolved into modern-day whales.

Some gaps in the fossil record still exist, but this can hardly be surprising. For an organism to be preserved as a fossil, a specific set of circumstances must occur quickly after death. After death, for example, the organism must be quickly covered with sediment to create an environment that is not conducive to decay. Even under ideal conditions, it is usually only the harder structures such as bones or feathers that tend to become mineralized. Soft tissues almost always decay and thus soft-bodied organisms are found much less frequently in the fossil record. Despite the relative rarity with which organisms become fossilized, the fossil record provides a remarkably complete picture of the evolution of many groups of organisms.

4. **The combination of nucleotide sequences in DNA requisite to produce a functional gene is too small to have been produced by random chance.**

   This criticism is based on the precision of gene sequences. Just one base-pair substitution out of tens of thousands of base pairs on a gene can render the gene nonfunctional. This criticism fails to understand two processes, however. First, the nucleotide (base-pair) sequences that make up a gene are under selection pressures that are not random. If a mutation results in a sequence that produces a nonfunctional gene, then the organism bearing that gene will likely die before reproducing, and that gene sequence will disappear from the population. The functional genes that we see in living organisms today possess a particular nucleotide sequence precisely because they have been selected for by environmental conditions. Genes lacking such a favorable sequence simply cease to exist. Further, just because the probability of a particular sequence existing is extremely low does not mean it cannot happen. Consider a room that contains 100 people. There are 365 days in a year; thus, each person has a 1 in 365 chance of being born on a particular day of the year. In a room of 100 people, the probability of everyone having their birthday on the sequence of days that they do is 1 in 5.9×10^{257} (that is, 5.9 followed by 257 zeros). That is a phenomenally small probability, yet there sits everyone with their birthday on those particular days.

5. **The intelligent design argument meets the criteria of sound scientific explanation.**

   Intelligent design (ID), despite claims to the contrary, is essentially a form of creationism that masquerades as science. Proponents of intelligent design argue that ID research is science and that its claims are testable as required by science; this is simply false. For example, the Discovery Institute claims that the search for evidence of a designer of life (i.e., God) is based on the principles of the scientific method. As an example, they offer that ID research begins with the observation that some intelligent agent produced complex information. Here ID fails at the first step of the scientific process: it is unclear what is meant by "complex information," and science requires direct, clear definitions of the problem at hand. Since ID spends much time discussing the complexity of living organisms, a safe assumption of the definition of "complex information" might be a living organism. No one has ever observed God or any other intelligent agent actually creating a living organism; thus, ID does not begin with the basic scientific step of observation. The so-called hypothesis that ID rests upon is that of irreducible complexity. This argument states that extremely complex structures could not have been produced by the slow, incremental changes of evolution. If one small part of a complex structure were missing, then the whole structure would not work as it is supposed to. As such, the structure could not have evolved through multiple small steps. This argument was first made regarding the vertebrate eye. It was stated that the eye is so complex it must have been created by a Designer all at once. Land and Nilsson[10] have shown quite elegantly how the eye could have evolved through incremental steps. In short, they have demonstrated that pigments that detect light (the earliest form of an eye) would have benefited early organisms. As additional complexity evolved, the eye would provide more and more benefit.

10. M. F. Land and D.-E. Nilsson, *Animal Eyes* (New York: Oxford University Press, 2002.)

Michael Behe, a professor of biochemistry at Lehigh University, is one of the rare scientists to support ID, arguing for irreducible complexity. One of Behe's claims against the theory of evolution is that the complex structure of the flagellum of single-celled organisms could not have evolved in incremental steps; if any component of the flagellum were missing, it could not move and propel the cell forward. Behe's argument regarding the flagellum fails because it assumes the components comprising a flagellum could not have been used for any other purpose in the cell. Evolution does not assemble complex structures in a single step. The early flagellum was almost certainly a much simpler structure and its individual components probably did not provide this single-celled organism with the same functions as in the flagellum of today. Instead, over evolutionary time, the structural components of the flagellum that were originally used for one purpose were later co-opted for a different purpose. The first flagellum almost certainly did not look like the fully formed structure it is today.

Supporters of ID claim they are testing the hypothesis of "irreducible complexity" by reverse engineering complex parts of a living organism. If the complex structure cannot function with a missing part, ID supporters claim that the structure must have been designed wholesale, rather than having evolved. Remember, however, that a scientific hypothesis must be both testable AND falsifiable. ID provides no falsifiable alternative to "irreducible complexity."

DISCUSSION QUESTIONS

1. Explain why an individual organism cannot evolve over time, but a population can.
2. State two requirements that a claim about the world must meet for the claim to be scientific.
3. Clarify the difference between a scientific *theory* and the word *theory* as it is used in everyday speech.
4. Is a scientific theory a powerful conceptual explanation of the world? Why or why not?
5. Certain groups often demand that creationism or intelligent design be taught in science class as an alternative to the theory

of evolution. Should creationism be taught as an equal alternative to evolution? Does creationism meet the criteria for scientific theory?

6. Define scientific knowledge. What are the limits of this knowledge? Is it possible to hold a worldview that accepts evolutionary theory and a reasoned faith?

GLOSSARY

adaptive radiation. The rapid (in evolutionary time) development of many new species from a common ancestral species. For example, the hundreds of cichlid fish species living in Africa's great rift lakes evolved from a common ancestral fish species within the span of a few million years.

allele. Different forms of the same gene. Different alleles perform the same function, but produce different variants. Genes that specify different eye color in humans are an example of this.

allopatric speciation. Evolution of two new species when the population of the original species is separated by a physical barrier, such as a mountain range.

gene. A segment on the DNA molecule that contains a specific code. This code tells the cell how to make a protein. Proteins are then used to create all the structures in an organism's body. Thus, genes are "blueprints" that tell cells how to manufacture all of the proteins needed to build and maintain the living organism. One DNA molecule often contains thousands of individual genes.

genotype. Sequence of genes possessed by an organism contributing to all of its characteristics.

inductive logic. A process of reasoning whereby a general conclusion is drawn from a limited set of evidence (i.e., making a broad generalization from a few specific examples).

microevolution. Small genetic changes occurring within a species over two or more generations. Microevolution often leads to macroevolution, the transition of one species into different species over evolutionary time.

mutation. A random change in the coding sequence of a gene. This can come about through some agent that damages a cell's DNA (e.g., a cancer-causing agent) or may occur purely by random chance when a cell makes an error when making a new copy of DNA during the process of cell division.

phenotype. Physical appearance of an organism produced by its genotype.

population. Interbreeding organisms (a particular species) located in a specific geographic region.

replication. The process by which a cell makes a new copy of DNA. For any organism to grow or reproduce, its cells must divide to make new cells. The copy (replicated DNA) is then donated to the new cell that is made during cell division.

species. A biological classification containing individuals that resemble one another and are capable of interbreeding. If two individual organisms cannot produce an offspring that is also capable of reproducing, then they are considered to be different species. This is one of the most commonly used definitions in biology. There are actually more than twenty different definitions used among biologists, indicating just how difficult it is to truly define a species. Because evolution is a gradual process, we should expect that species boundaries will not always be clearly defined.

RESOURCES FOR FURTHER STUDY

Andersson, M. "Female Choice Selects for Extreme Tail Length in a Widowbird." *Nature 299* (1982): 818–20.

Andersson, M. *Sexual Selection*. Princeton, NJ: Princeton University Press, 1994.

Basolo, A. "Female Preference Predates the Evolution of the Sword in Swordtail Fish." *Science* 250 (1990): 808–10.

Darwin, C. *On the Origin of Species by Means of Natural Selection, or The Preservation of Favoured Races in the Struggle for Life*. London: J. Murray, 1859.

Darwin, C. *The Descent of Man and Selection in Relation to Sex*. London: J. Murray, 1871.

Futuyma, D. J. *Evolutionary Biology.* Sunderland, MA: Sinauer Associates, 1997.

Gerhardt, H. C., and F. Huber. *Acoustic Communication in Insects and Anurans: Common Problems and Diverse Solutions.* Chicago: University of Chicago Press, 2002.

Gross, M. R. "Sneakers, Satellites, and Parental: Polymorphic Mating Strategies in North American Sunfishes." *Z. Tierpsychol* 60 (1982): 1–26.

Hebets, E. A., and G. W. Uetz. "Leg Ornamentation and the Efficacy of Courtship Display in Four Species of Wolf Spider (*Araneae: Lycosidae*)." *Behavioral Ecology and Sociobiology* 47 (2000): 280–86.

Kodric-Brown, A., and J. H. Brown. "Truth in Advertising: The Kinds of Traits Favoured by Sexual Selection." *American Naturalist* 124 (1984): 309–23.

Land, M. F., and D.-E. Nilsson. *Animal Eyes.* New York: Oxford University Press, 2002.

Petrie, M., T. Halliday, and C. Sanders. "Peahens Prefer Peacocks with Elaborate Trains." *Animal Behavior* 41 (1991): 323–31.

Ridley, M. *Evolution.* Oxford: Blackwell Pub., 2003.

Ryan, M. J. *The Túngara Frog: A Study in Sexual Selection and Communication.* Chicago: University of Chicago Press, 1985.

Ryan, M. J., ed. *Anuran Communication.* Washington, DC: Smithsonian Institution Press, 2001.

Scott, E. C. *Evolution vs. Creationism: An Introduction.* Berkeley: University of California Press, 2004.

chapter 3

From Exception to Exemplification: Understanding the Debate over Darwin

Brian G. Henning
Gonzaga University, Spokane, WA

Although the contemporary debate over evolution began in 1809, which may seem like ancient history to many, the careful student of history will know this debate has a much more ancient pedigree. Indeed, in many ways, the debate over evolution began some 2,400 years ago, in ancient Greece.

ATOMS OR ESSENCES?

Democritus (460–370 BCE), perhaps the most prolific of philosophers born before Socrates (469–399 BCE), is often seen as a herald of modern science.[1] What is most pertinent here, in discussing issues related to the nature of life, is Democritus's theory of **atoms**, a theory that led him to become a **determinist** and a **materialist**.

Determinism posits that every event in nature is determined by absolute laws of nature; thus, according to this view, there is no genuine novelty, much less freedom. This view, in turn, is frequently the result of a materialist notion of reality. While today to be a materialist suggests one cares only about nice clothes and money, in its philosophical sense, to be a materialist means one believes reality is composed solely of matter, or atoms. Materialism thus denies the possibility of all nonmaterial forms of reality (e.g., a nonmaterial

1. Pamela Gossin, *Encyclopedia of Literature and Science* (Westport, CT: Greenwood Press, 2002).

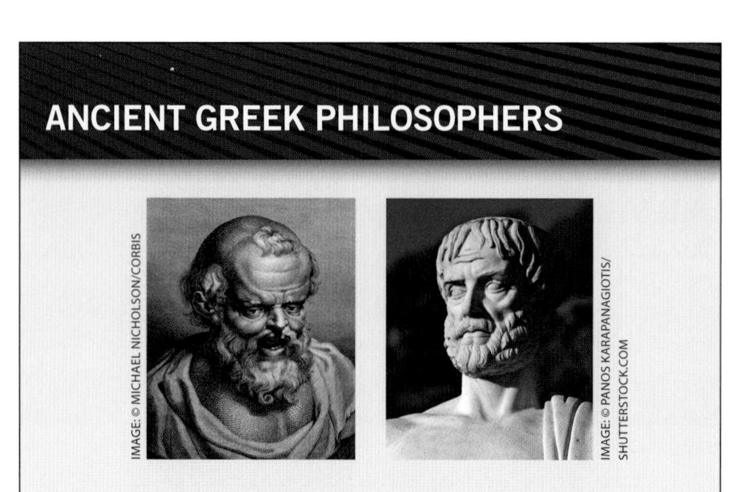

Democritus (460–370 BCE), sometimes referred to as the father of modern science, developed a theory of atoms.

Aristotle (384–322 BCE), Plato's most famous student, went on to found his own school and is considered the first biologist.

mind or soul or a transcendent God). For the ancient Greeks, like Democritus, atoms were deemed the indivisible building blocks from which everything is made.[2] Democritus's view of nature—nature as composed of atomic building blocks controlled by absolute laws of nature—is surprisingly similar to the underlying worldview of many contemporary scientists, which will be explained more fully later.

Aristotle (384–322 BCE), the son of a doctor attached to the Macedonian court in what is present-day northern Greece, rejected the idea that nature can be understood in terms of matter alone and argued in part that Democritus' view does not account for the origination of motion. Whereas Democritus believed every event to be completely determined by prior events, Aristotle believed the order of nature must be explained in terms of purpose; he sought teleological or goal-oriented answers as opposed to mechanistic ones.[3] For

2. The Greek word *atomos* or *a-tomos* literally means "not cuttable" or "indivisible."

3. The Greek word *telos* translates as "end" or "purpose."

Aristotle, nature is not composed exclusively of matter, but rather of matter and form, body and soul. A plant or an animal, for instance, has the characteristics and powers that it does—such as growth, reproduction, and metabolism—because it has a ***psychē*** (pronounced soo-khay), an internal principle of organization and change that arranges its matter.[4] Aristotle argued that it is this internal, invisible form and not absolute laws of nature that gives matter its capacities and molds it into one particular individual as opposed to another. According to this view, an individual plant or animal is not merely a complex machine acting out its preset programming, as it was for Democritus and indeed as it continues to be for many contemporary scientists. For Aristotle, each individual in nature is a complex composite of form and matter that acts in order to achieve what is best for it. In this view, the order of nature is neither blind nor necessary; everything in nature has a purpose and acts as it does because it is good.

To better understand Aristotle's claim that everything in nature happens because it is good, take the example of why rain falls. For Democritus and the atomists, rain falls because water that has evaporated and been "drawn up" (Aristotle's words) eventually cools and, once cool, again becomes water and falls as rain. The fact that rain waters crops is seen as purely coincidental by the atomists, a view that particularly bothered Aristotle. For Aristotle, rain falls, not by necessity or chance, but because it is good (in this case, good for crops; this is its purpose, its ***telos***).[5] After considering the example of rain, Aristotle further illustrates his point by considering the origins of parts of animals. In the process, Aristotle considers, for the first time in the history of Western thought, the idea that natural selection could be based on fitness for survival.

> Why then should it not be the same with the parts of nature, e.g., that our teeth should come up of necessity—the front

4. I have avoided translating the Greek word *psychē* as "soul" because the connotations of *soul* are unavoidably religious and, unlike the philosophical tradition, influenced by two thousand years of Christian history.

5. Aristotle, *The Complete Works of Aristotle*, vol. 1, *Physics*, ed. Jonathan Barnes (Princeton: Princeton University Press, 1984), 198b15–32. (Note that references to Aristotle's works use "Bekker numbers" or standard marginal letters and numbers, not page numbers.)

> teeth sharp, fitted for tearing, the molars broad and useful for grinding down the food—since they did not arise for this end, but it was merely a coincidental result; and so with all other parts in which we suppose that there is purpose? Wherever then all the parts came about just what they would have been if they had come to be for an end, such things *survived*, being organized spontaneously in a *fitting* way; whereas those which grew otherwise perished and continue to perish, as Empedocles says his "man-faced ox progeny" did.[6]

But for Aristotle, front teeth are sharp *not* because, having arisen by chance, they were more "fit" and "survived." Indeed, he argues that "it is impossible" that the order of nature is the "result of chance or spontaneity." Rather, rain falls and teeth are sharp, not by chance, but "for the sake of something." Rain falls because it is good for crops and front teeth are sharp because they are good for the animal. Incredibly, Aristotle considers and then rejects the concept of survival of the fittest.

It is important to note that, for Aristotle, the origin of each species was not considered a problem in need of a solution, because Aristotle believed every living species had always existed. Indeed, the origin of the universe itself was not a problem for Aristotle, because he believed there was no start to the universe; it has always been, so there is no need to account for its origin.

The view of nature as purposive committed Aristotle to a powerful, hierarchical conception of nature. The purpose of each living thing is not merely tied to what is best for it, each living thing also has a particular place in the larger order of nature. Thus, for example, in addition to immanent aims such as growth, nourishment, and reproduction, plants' purpose is "for the sake of" animals, and animals' purpose is "for the sake of" humans (see figure 1). For instance, in Aristotle's *Politics*, he argues that "after the birth of animals, plants exist for their sake, and that the other animals exist for the sake of man. . . . Now if nature makes nothing incomplete, and nothing in vain, the inference must be that she has made all animals for the sake

6. Ibid., emphasis added.

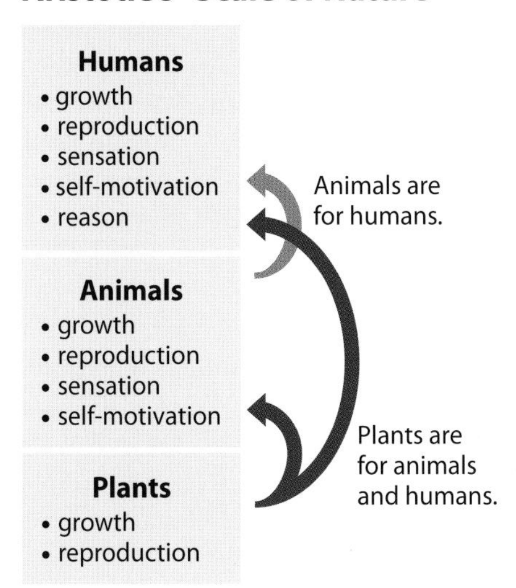

Figure 1. Author illustration

of man."[7] In many ways, this hierarchical notion of reality—what is sometimes referred to as the *scala naturae* or "scale of nature"—seems to capture something important about the differences between humans and other animals. This view of reality expresses a deeply held intuition, avowed by philosophers and theologians for centuries, regarding the superiority of humans over the rest of nature.

However, it is important to note that this hierarchy held not only *between* species but also *within* species. "And so," Aristotle continues, "from one point of view, the art of war is a natural art of acquisition, . . . an art which we ought to practice . . . against men who, though intended by nature to be governed, will not submit; for war of such a kind is naturally just."[8] Not only do plants exist for animals and animals for humans, but "inferior peoples," particularly the Germanic tribes in the north, exist for the sake of the "superior" Greeks, according to Aristotle. Inferior people ought, therefore, to

7. Aristotle, *The Complete Works of Aristotle*, vol. 2, *Politics* I.8., ed. Jonathan Barnes (Princeton: Princeton University Press, 1984).

8. Ibid., 1256b15–25.

submit to the Greeks because, being superior, the Greeks are "by nature" intended to govern the inferior. Should the inferior peoples refuse to recognize this, the Greeks then considered a war to subjugate them as naturally just. In this way, Aristotle's hierarchical view of nature has also been used in modern times to justify the enslaving of peoples perceived as inferior, the often violent and coercive "reeducation" of indigenous peoples, the subjugation of women, the cruel use of nonhuman animals for often trivial purposes, and the wanton destruction of the natural world.[9] It will be important to return to this point.

In the end, Aristotle's view of nature triumphed over Democritus's atomistic theory. In the West, Aristotle's view of reality was taken as the undisputed truth for nearly two millennia. During the so-called Islamic Golden Age,[10] Arabic-speaking philosophers in northern Africa and the eastern Mediterranean held Aristotle in such high regard they simply called him "The Philosopher." Thomas Aquinas (AD 1225–1274), writing at the end of the Middle Ages, held Aristotle in similar esteem, devoting many of the thousands of pages of his work to a grand synthesis between Aristotelian and Christian worldviews.[11]

Indeed, Aristotle's basic view of nature was virtually unchallenged for nearly 1,800 years. It was not until the advent of the modern era and in particular the rise of modern science—with its emphasis on the categorization of known facts and the inductive discovery of new ones—that Aristotle's hierarchical and teleological idea of nature was challenged. As social changes in Europe gradually created intellectual space, scholars began to embark on a new project, marked by the explicit repudiation of the Aristotelian worldview.

---

9. For instance, Aristotle's hierarchical thinking led to the view that some people are "natural slaves," who by nature are "living tools" to be used by "freemen." Further, Aristotle explicitly argued that women were inferior to men and, indeed, were incomplete or deformed men. For a catalog of these and other Aristotelian claims, see Cynthia Freeland, "Nourishing Speculation: A Feminist Reading of Aristotelian Science," in *Engendering Origins: Critical Feminist Readings in Plato and Aristotle*, ed. Bat-Ami Bar On (Albany: State University of New York Press, 1994): 145–46.

10. The Golden Age of the Islamic world is often dated from the mid-eighth century to the mid-thirteenth century.

11. Aristotle's texts were lost to Western Europe until the late Middle Ages (twelfth century), though this region was essentially Aristotelian in much of its outlook.

THE RISE OF MODERN SCIENCE

Philosophically, the modern era (1500–1800) is characterized by a clear rejection of the Aristotelian worldview, particularly its teleological view of nature with its hidden forms or essences. Scholars were increasingly frustrated by the intractability of the philosophical debates in the late Middle Ages and by the general failure of philosophy to arrive at any lasting achievements. Meanwhile, new methods in science and mathematics were yielding impressive results. It was only a matter of time before philosophers sought to model their investigations on these more successful approaches.

Francis Bacon (1561–1626) was among the earliest and most vocal opponents of the Aristotelian model of the late Middle Ages. Bacon was relentless in his attacks on the Aristotelian worldview, with its esoteric truths and invisible essences.[12] For Bacon, science, if it was to be of any use at all, should be aimed at improving people's lives. Thus, he proposed a new method, focused not on the deductive categorization of natural types, but on the inductive discovery of nature's order through active experimentation. Bacon is rightly credited as one of the earliest and most vocal proponents of what is now called the scientific method. For Bacon, the primary purpose of scientific investigation is not to discover invisible forms in nature, as Aristotle suggested, but to uncover, through inductive experimentation, the laws of nature. True to his method, Bacon is reported to have died from pneumonia, contracted while testing a hypothesis about the preservation of flesh (he is reported to have been stuffing a dead chicken full of snow).

Although boundaries between eras are never sharply defined, many rightly suggest the modern era truly began with the Frenchman Rene Descartes (1596–1650). Having made impressive, original contributions in mathematics,[13] Descartes sought to model philosophy on the methods of geometry. As with a geometrical proof,

---

12. Although Aristotle rejected the rationalism of Plato and Socrates, which preached the mistrust of the senses, Aristotle's empiricism differs from the modern scientific method. For Aristotle, scientific investigation was primarily an act of categorization, not discovery; thus, there is no true novelty in nature. In a sense, Aristotle's method was empirical, but it was not experimental.

13. Elementary students of geometry have Descartes to thank (or blame) for modern analytic geometry, which was once referred to as Cartesian geometry, after Descartes.

1791 This engraving is by Marie Champion de Cernel after a drawing by Antoine-Francois Sergent-Marceau.

Descartes thought the first step in philosophy should be to break down problems into their smallest parts. For Descartes, this meant one must question everything to determine if there is anything at all that can be known with certainty. Barring some firm and indubitable foundation, he thought, philosophy will forever fail to make real contributions to understanding. In his famed *Meditations* (1641), Descartes "discovers" this firm foundation in the realization that, while it is possible to doubt everything else, including the existence of the physical world and his own body, it is impossible to doubt his own existence, for, in so doing, he only establishes it more firmly. Thus, we have Descartes' renowned dictum, "*cogito, ergo sum*" or "I think, therefore I am."

What is relevant here is the effect Descartes' philosophy had on how humans and the natural world are understood. Descartes replaced the Aristotelian notion of individuals as complex beings made of form and matter and replaced it with a **dualistic** view of nature, or the view that there are only two kinds of substances in the universe: mind and matter. In a sense, Descartes takes the Aristotelian scale of nature with its complex vertical structure and lays it on its side, making only one cut (see figure 2). On one side, there are those beings composed of matter, which are mechanistically determined by the "disposition of their organs"[14] or the arrangement of their parts; on the other side, there are beings that can think and are therefore free. In creating this unbridgeable chasm between mind and matter,

14. Rene Descartes, *The Philosophical Writings of Descartes*, vol. 1, *Discourse on the Method*, trans. John Cottingham, Robert Stoothoff, and Dugald Murdoch (Cambridge: Cambridge University Press, 1985), 141.

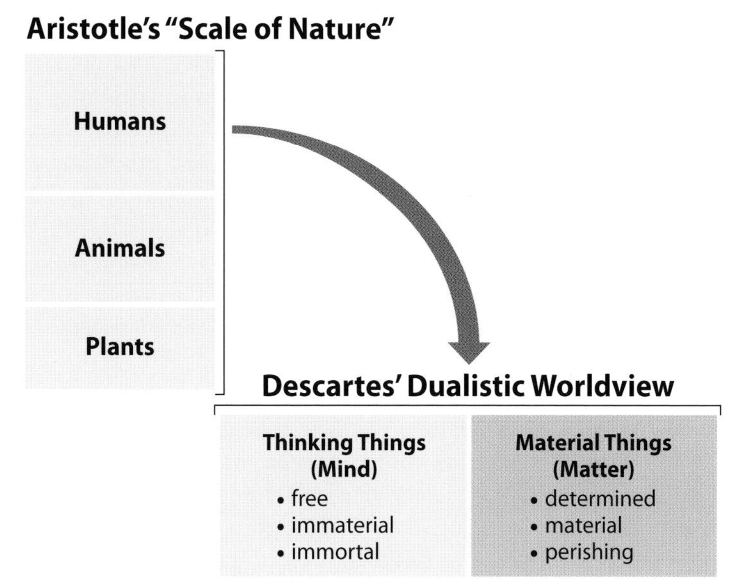

Figure 2. Author illustration

Descartes' dualistic worldview fundamentally alters the conception of human beings and the natural world as follows.

First, according to Descartes' dualism, I as a human being am not my body. The one thing of which I can be certain, Descartes writes, is that I am nothing but this immaterial "thinking thing" that is called mind. Aristotle and the ancient Greeks held that humans were unique in their capacity for reason; likewise, medieval philosophers, adding to this conception, held that humans were uniquely made in God's image. But Cartesian dualism takes this separation between humankind and all other living things to a new level. While, for the Greeks and medieval philosophers, humans were set at the top of nature's hierarchy, humans were nonetheless a fundamental part of nature or the created order. By embracing a sharp and fundamental distinction between beings in nature, which are composed entirely of matter, and human beings, who alone have a mind (or more precisely *are* a mind), Descartes essentially removes humans from the natural order. Human beings are a unique and singular exception; humans are the only free beings in an otherwise clockwork universe.

Descartes' reduction of nature to mechanistically determined bits of matter is equally dramatic in its effects. Descartes declares an absolute chasm between free, mental substances and determined, material substances: it is not that "beasts have less reason than men, but that they have no reason at all."[15] Animals and other natural beings are like sophisticated clocks whose inner workings—having been made by the infinitely skilled Craftsman (God)—allow their complex behaviors.[16] This reductive conception of nature, which increasingly defined the modern understanding of the natural world, has had and, in a sense, continues to have, devastating implications for how human beings treat animals and plants.

Joined with the long-standing notion of our biblically prescribed dominion over nature, the view that animals are complex machines that neither think nor feel pain or pleasure has made possible some of the more violent abuses of animals and the natural world.[17] The consistency with which Descartes himself held this view is noteworthy. For instance, in his day, the circulation of the blood was a subject of great debate. Arguing strenuously against the English physician William Harvey's[18] view that the heart acted as a pump, Descartes claimed instead that blood circulated through the body due to heat differentials. He sought to demonstrate this through **vivisection**, or live dissection. In one work, he describes a "striking experiment" whereby "If you slice off the pointed end of the heart in a live dog, and insert your finger into one of the cavities, you will feel unmistakably every time the heart gets shorter it presses the finger, and every time it gets longer it stops pressing it."[19]

---

15. Ibid., 58, 140.

16. Cf. "It proves rather that they have no intelligence at all, and that it is nature which acts in them according to the disposition of their organs. In the same way a clock, consisting of wheels and springs, can count the hours and measure time more accurately than we can with all our wisdom" (ibid., 141).

17. For a more developed treatment of this claim, see Brian G. Henning, "From Despot to Steward: The Greening of Catholic Social Teaching," in *The Heart of Catholic Social Teaching: Its Origins and Contemporary Significance*, ed. David Matzko McCarthy (Grand Rapids, MI: Brazos Press, 2009), 183–93.

18. The English physician William Harvey (1578–1657) is credited with first describing the proper functioning of the circulatory system, particularly the role of the heart in the circulation of blood.

19. Descartes, *The Philosophical Writings of Descartes*, vol. 1, *Description of the Human Body and All of Its Functions*, trans. John Cottingham, Robert Stoothoff, and Dugald Murdoch (Cambridge: Cambridge University Press, 1985), 317.

Similarly, in a letter to a colleague, Descartes describes another experiment in which he "opened the chest of a live rabbit and removed the ribs to expose the heart." After he "cut away half the heart," he noticed that the separated part stopped beating instantly, while the part still attached "continued to pulsate for quite some time."[20] While the contemporary reader is likely to be sickened by such treatment, Descartes was simply acting consistently with his view of reality. Lacking a mind with which to understand and hence to feel, animals are nothing more than complex machines, he believed. As one account put it:

> The (Cartesian) scientists administered beatings to dogs with perfect indifference and made fun of those who pitied the creatures as if they felt pain. They said the animals were clocks; that the cries they emitted when struck were only the noise of a little spring that had been touched, but that the whole body was without feeling. They nailed the poor animals up on boards by their four paws to vivisect them to see the circulation of the blood which was a great subject of controversy.[21]

Logically, if animals are merely complex machines, then one is justified in treating them as one would any other machine. What one does with and to a machine is a matter of prudence, not morality. The act of cutting open the chest of a live dog and manipulating its heart is not fundamentally different than opening the hood of one's car and manipulating the hoses and clamps that circulate the oil. The sound one hears coming from a living dog with its chest cut open is not the result of pain, because, lacking a mind, the dog has no such experiences. Rather, Descartes would suggest, the sounds are the natural result of a machine that is under strain, just as one would hear sounds if one dropped a wrench in a running car engine; the sound is terrible, but one ought not to infer the car feels pain. Though animals are perhaps more complex in the arrangement of

---

20. Descartes, "To Plempius, 15 February 1638," in *The Philosophical Writings of Descartes*, 3:81–82, trans. John Cottingham, Robert Stoothoff, Dugald Murdoch, and Anthony Kenny (Cambridge: Cambridge University Press, 1991), 81-82.

21. Leonora Cohen Rosenfield, *From Beast-Machine to Man-Machine: Animal Soul in French Letters from Descartes to La Mettrie* (New York: Octagon Books, 1968), 54.

their parts, Descartes suggests, they are not fundamentally different from any human-made machine.

The depiction of nonhuman nature as a complex machine composed of material parts and organized by natural laws has proved a wildly successful research paradigm. Combined with the self-correcting nature of an experimental, inductive method, science has gradually begun to lay bare blueprints of the universe. Yet, throughout the modern era, a basic incoherence nagged at the consistency of the mechanistic worldview. While some, such as the Englishman David Hume (1711–1776), simply dispatched entirely the notion of an immaterial substance called mind or soul and embraced full materialism, many others, amid the humanistic fervor of the Enlightenment, sought to retain a special place for the human separate from the rest of the natural order.[22] It was not until the adventures of another Englishman, Charles Darwin (1809–1882), that a coherent conception of the origin of human beings was achieved.

OUR EVOLUTIONARY NEXT OF KIN

With the publishing of *On the Origin of Species by Means of Natural Selection* (1859), Darwin fundamentally challenged previous ideas about the origins of life on Earth. Darwin found that, contrary to Aristotle's claim, some teeth *are* flat because, by chance, they made a particular individual more fit, allowing it to survive longer, reproduce more, and pass on its traits. Each organism has evolved over millions of generations to survive in a certain ecological niche. Thus, it is no more correct to say that plants exist for the sake of animals than animals exist for the sake of plants. Each species possesses its particular traits because those traits, arising by chance mutations, provided adaptive advantages. As the respected Georgetown theologian John F. Haught notes, Darwin's findings rendered the old hierarchical notion of nature untenable.

22. The most notable example of the latter is the German philosopher Immanuel Kant (1724–1804), who was woken from his "dogmatic slumbers" by the controversial views of David Hume.

> Evolutionary science has blurred the former lines between nonlife, life, humanity, and culture to the point where it is hard to decide clearly where one begins and the other leaves off. Evolutionary science posits a physical and historical *continuity* running through all those levels of nature formerly thought of as discontinuous and hierarchically distinct. In all of science nothing seems to have melted down the classical hierarchical vision more completely than has the evolutionary picture of nature. In combination with physics, chemistry, molecular biology, geology, and other sciences, neo-Darwinism has now made the traditional idea that nature leads upward to God by way of a series of hierarchically distinct rungs on the cosmic ladder seem quite unbelievable.[23]

With the publishing of *The Descent of Man, and Selection in Relation to Sex* (1871), Darwin fundamentally challenged how humans view their place in the cosmic order. Humans were no longer categorically set at the pinnacle of, or apart from, nature (though humans historically have been reluctant to give up their **anthropocentrism**, even while intellectually assenting to an evolutionary worldview).[24] For millennia, philosophers and theologians had depicted humans as utterly unique, but Darwin reveals this as a great delusion; humans have been viewing nature from a self-constructed pedestal. According to Darwin, evolutionary change occurs gradually. Thus, if humans have

Charles Darwin

---

23. John F. Haught, *God After Darwin: A Theology of Evolution* (Boulder, CO: Westview Press, 2008), 64.

24. *Anthropocentrism* literally means "human-centered." It is the view that all meaning and value is derived from an entity's relationship to human beings; an entity has no value outside its relationship to human beings.

humor, emotions, language, and a highly developed capacity for reason and culture-making, among other traits, then our evolutionary ancestors must also have had similar, though perhaps less acute, capacities. Evolutionary theory requires that one abandon simplistic, binary conceptions. Language, reason, emotion, and culture are not all-or-nothing categories; rather, these traits are possessed by many living creatures to varying degrees.

In his engaging and entertaining book *Next of Kin*, Roger Fouts provides powerful evidence for this claim. While working his way through graduate school, Fouts took a job as a lab assistant for two psychologists who were exploring whether it was possible to teach a chimpanzee named Washoe a human language, American Sign

Washoe is seen in a 1995 handout photo from Central Washington University. Washoe, the chimpanzee researchers said was the first non-human to acquire a human language, died Nov. 1, 2007, after a brief illness. She was 42. Washoe first learned American Sign Language in a research project in Nevada, according to the Chimpanzee and Human Communication Institute (CHCI) where she had lived since 1980.

Language (ASL). This unintentional encounter started what would become a lifelong relationship between Fouts and Washoe. Washoe learned, not by memorizing and mimicking, but by participating in a language community.[25] Washoe was able to understand hundreds of signs and could combine them to communicate in syntactically correct sentences.

> Her performance, at four years of age, was remarkable. She scored 86 percent correct in one representative test that had 64 trials, and on a test twice as long—128 trials—she scored 71 percent. (Guessing randomly on this test would produce a score of 4 percent) . . . [A]t age five, Washoe was using 132 signs reliably and could understand hundreds of others. In addition to naming and categorizing objects, Washoe began doing something that [Noam] Chomsky said only humans could do: she assembled words into novel combinations.[26]

Perhaps even more striking, in a later stage of the experiment, without any influence from the human staff, Washoe taught ASL to her adopted son, Loulis. Fouts chronicles story after story demonstrating that Washoe made jokes, could understand some complex concepts, planned for the future, and nearly died from severe depression caused by the death of her newborn son.[27] "The apparently intelligent chimpanzee now threatened to bring down the entire artifice of Aristotle's Great Chain of Being."[28]

Fouts's point was not that there was no difference between himself and Washoe or between humans and chimpanzees. Important differences in the abilities of chimps and humans exist, but these differences are ultimately a matter of degree, not kind. To be more precise, differences of kind are a consequence of differences of degree.

---

25. Roger Fouts and Stephen Tukel Mills, *Next of Kin: My Conversations with Chimpanzees* (New York: Avon Books, 1997), 77f.

26. Ibid., 100–101.

27. Washoe's son Sequoyah died on March 8, 1979.

28. Fouts, *Next of Kin*, 50.

To understand this claim regarding differences, imagine the colors red, orange, and yellow on the color spectrum.

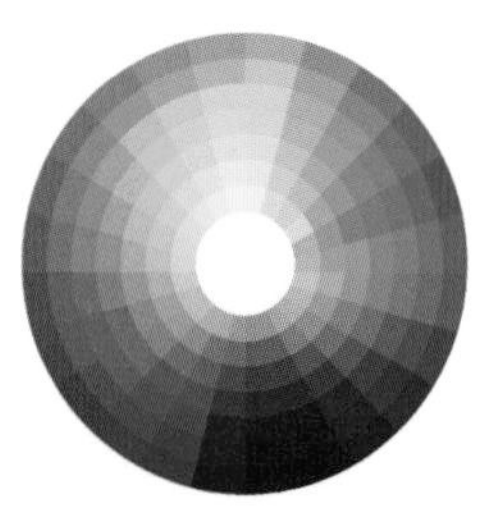

IMAGE: © EKLER/SHUTTERSTOCK.COM

While it is undeniable that red, orange, and yellow are each a different *kind* of color, the differences between these distinct colors are ultimately a result of differences of *degree*. Minute changes in the intensity of the color red gradually shift to become orange, which gradually becomes yellow, and so on. Notice that *between* these colors it is impossible to select one particular point and say, "Here, this is clearly only red and not orange," or "This is clearly only orange and not yellow." Rather, there is orange-yellow and then yellow-orange and so on.[29] There is no clear demarcation. Yet, this does not deny the fact that red, orange, and yellow are distinctly different colors. When differences of degree accumulate sufficiently, they result in differences of kind.

In the same way, evolutionary theory teaches us that each species is a product of gradual accumulations of minute changes over millions of years, changes which result in distinct species of animals, such as chimps and humans. Chimps are clearly one kind of animal and humans another, but the differences between us are ultimately only a matter of degree. In an important way, this is what the experiment with Washoe is meant to demonstrate. If chimpanzees are indeed our closest evolutionary relatives, sharing 99.4 percent of our DNA,[30]

29. Crayon manufacturers are creative in picking names to reflect these subtle color changes, referring to yellow, and also daisy yellow, goldenrod, and sunset.

30. Having branched off from humans more recently (perhaps 5.4 million years ago), chimps are genetically more similar to humans (0.6 percent difference in DNA) than to gorillas (2.3 percent difference) or orangutans (3.6 percent difference) (Fouts, *Next of Kin*, 55). "Despite all outward appearances, the chimpanzee's next of kin is not the gorilla or the orangutan but the human" (ibid.).

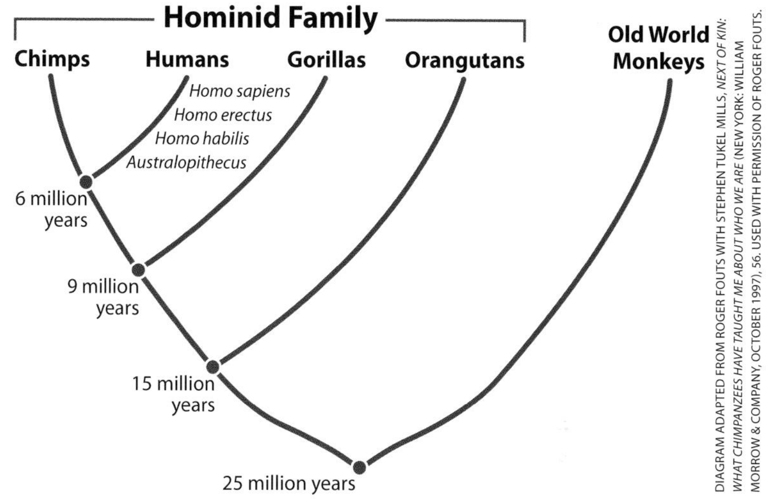

DIAGRAM ADAPTED FROM ROGER FOUTS WITH STEPHEN TUKEL MILLS, *NEXT OF KIN: WHAT CHIMPANZEES HAVE TAUGHT ME ABOUT WHO WE ARE* (NEW YORK: WILLIAM MORROW & COMPANY, OCTOBER 1997), 56. USED WITH PERMISSION OF ROGER FOUTS.

Figure 3. The numbers in this diagram indicate the estimated millions of years from the last branch point to the present along particular lineages of humankind's evolutionary tree.

then, Fouts reasoned, they ought to be capable of complex thought and language.

This is exactly what he found. Properly understood, the theory of evolution leads us to question simplistic claims of human uniqueness, claims that humans alone are capable of consciousness, self-awareness, reason, language, and emotion. If we have evolved over tens of millions of years from other life-forms, then, of necessity, our evolutionary ancestors possess our own traits in simpler forms.

BUT WHY AREN'T WE ZOMBIES?

Darwin's theory of evolution via natural selection rightly challenges the arrogant anthropocentrism or human-centered attitude that has been the basis of Western self-definition. But neo-Darwinism, which combines genetic theory with Darwin's theory of natural selection, fails to question the modern understanding of the universe as a vast machine composed only of material constituents. Indeed, it begins with this assumption. As the Australian biologist Charles Birch puts it, "Neo-Darwinism, like all biological theories, is strictly mechanistic, no more and no less than the theory of DNA. Its entities, be they genes or organisms, are treated as objects and not as subjects. The dominant framework in which biologists discuss evolution is mechanistic and materialistic."[31] According to this "modern synthesis," organisms are merely "vehicles" for genes, which are the true objects of study.[32]

Birch has a memorable way of exposing the inadequacy of a completely mechanistic worldview. If the world is completely composed of inert bits of matter mechanistically determined by natural laws, he asks, then why are we not zombies? Zombies, he notes, are fictitious creatures identical to humans in every way except they are "devoid of any conscious experience."[33] If we evolved from simpler organisms, and those simple organisms are completely reducible to their material constituents, then it would stand to reason that we would similarly be devoid of conscious experience. You can't give what you don't have. Thus, inverting the question, we can realize that if humans have conscious experience and we evolved from simpler organisms, then simpler organisms must also have some form of conscious experience.

> If it is acknowledged that it [conscious experience] has evolved from animal experience, the question is pushed further back. At some point, does experience, mentality, or subjectivity emerge from entities that are totally lacking in these properties, entities that are simply objects for other subjects? If the basic constituents of the physical world are purely material, then the answer must be yes. But to believe this is to affirm a miracle. Unless you believe in miracles, you would expect mindless atoms to evolve into zombies. But this does not seem to happen.[34]

---

31. Charles Birch, "Why Aren't We Zombies?" in *Back to Darwin: A Richer Account of Evolution*, ed. John B. Cobb, Jr. (Cambridge: Eerdmans, 2008), 250.

32. To describe this worldview, Ian Barbour coins the memorable term *mechanomorphism* or the view that creatures are just like machines. Barbour, "Evolution and Process Thought," ibid., 213.

33. Birch, "Why Aren't We Zombies?" 251.

34. Ibid., 253. Birch further states, "One line of investigation of self-organization is to try to make computer models of its processes. These are necessarily mechanistic models. Some of the processes can indeed be replicated on a computer, but this does not mean that the behaving entities are in all respects like machines. They may partake in something like the purpose-motivated creativity of humans. Determinism by genes

What is missing from neo-Darwinism's mechanistic picture, Birch continues, is

> an understanding of what it is to be a living subject. Life is bound up with an urge to live. This principle is far more basic to life than the principle of survival of the fittest. The triumph of neo-Darwinism is within limits. Life is more than mechanics. Mechanistic procedures are entirely justifiable provided there is recognition of the limitations involved. What they ignore is experience, some of which at the human level is conscious. Experience is the inner aspect of entities. It has to do with feeling, valuing, purposing, and deciding.[35]

There is good reason, then, to question some of the assumptions of the dominant view of evolutionary theory.

While a powerful explanatory metaphor, conceiving of individuals as complex machines does not adequately characterize the nature of reality. Animals are not clocks, as Descartes suggested, nor are they computers, as many contemporary scientists seem to imply. As the British philosopher and mathematician Alfred North Whitehead (1861–1947) rightly noted at the beginning of the last century, "The only way of mitigating mechanism is by the discovery that it is not mechanism."[36] One must avoid what the eminent biologist Francisco Ayala calls the "nothing but" fallacy. One should not maintain that organisms are "nothing but" vehicles for genes. Indeed, as theologian and philosopher John Cobb, Jr., asserts, there is sound reason to claim that

> a materialistic treatment of evolution is profoundly inadequate and misleading . . . It is far closer to the facts to envision a field of events in which significant patterns can be discerned. The events are much better conceived in

---

is not an all-or-none affair. There can be different degrees of freedom. There is all the difference in the world between 100 percent determination and 99 percent determination. One provides no room for choice and purpose. The other does" (260).

35. Ibid., 253.

36. Alfred North Whitehead, *Science and the Modern World* (New York: Free Press, 1925), 76.

> organismic than in mechanistic terms. They are what they are in and through their relations with one another.[37]

Unlike machines, organisms are integrated wholes that are more than the sum of their parts. Whereas machines are limited to their "programming," organisms are capable of genuine, albeit limited, novelty. As the groundbreaking work of biophysiologist J. Scott Turner demonstrates, modern Darwinism is unable to account for this creative impulse, this intentional nature of organisms.

> On principle, modern Darwinism rules out any role for intentionality in our thinking about evolution. Evolution is immediate, contingent, and does not look forward. . . . Yet, we know that intentional living systems have evolved on Earth, because we are examples of them. We are capable of looking forward, assessing the future, and intentionally seeking future goals. How, then, can an unintentional process, which natural selection is supposed to be, produce intentional beings like ourselves? Was it just a lucky break, or is it a reflection of an unappreciated intentionality in the process of evolution? Modern evolutionary biology doesn't really have a good answer to this question.[38]

That we are not zombies, but intentional, creative agents implies that the potential for novel interaction and subjectivity reaches down to the most basic levels of reality.

FROM EXCEPTION TO EXEMPLIFICATION

To truly understand the significance of evolution, it is important to situate biological evolution within the wider scope of cosmic evolution, which began more than 13.6 billion years ago. Earth itself did not even arrive on the cosmic scene until 4.5 billion years ago. Our species, *Homo sapiens*, is a relative latecomer, appearing

---

37. John Cobb, Jr., "Organisms as Agents in Evolution," in *Back to Darwin: A Richer Account of Evolution*, ed. John B. Cobb, Jr. (Cambridge: Eerdmans, 2008), 241.

38. J. Scott Turner, *The Tinkerer's Accomplice: How Design Emerges from Life Itself* (Cambridge, MA: Harvard University Press: 2007), 29.

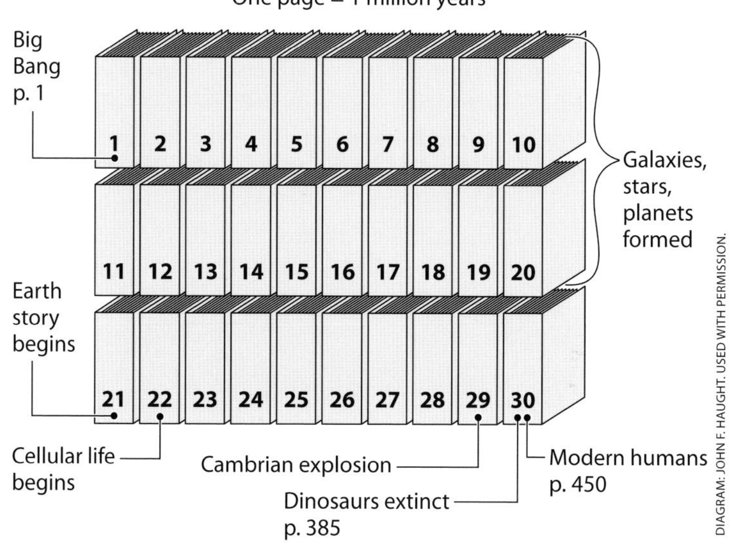

Figure 4. Volumes of books to represent cosmic evolution.

only 200,000 years ago. These vast stretches of cosmic time—what Haught calls "deep time"—are nearly impossible for humans to comprehend. Haught provides an illustration helpful in representing this concept.

Imagine time represented as a series of books in which each page represents 1 million years and each volume has 450 pages. The total history of the universe would be represented by thirty volumes. The first twenty volumes of the history of the universe contain essentially lifeless chemical processes that formed the first galaxies, stars, and terrestrial bodies. The story of Earth does not even begin until volume 21. Primitive, single-celled life-forms arrive on the scene in volume 22. The so-called Cambrian explosion, a period of rapid speciation during which most major animal groups are believed to have developed, occurs at the end of volume 29 (550 million years ago), and the dinosaurs go extinct on page 385 of volume 30 (65 million years ago). *Homo sapiens* do not appear until two-thirds of the way down the last page of volume 30. Recorded human history begins at the start of the last sentence of the last page of volume 30. Looking forward, the Sun is expected to have enough fuel to burn for another ten or eleven volumes (5 billion years) before it swells into a red giant, swallowing Earth and many other planets in the process.

Recognizing the great age of the universe is important in part because it makes an arrogant anthropocentrism more difficult to maintain. Recognizing the awesome length of the cosmic story and that Earth was teaming with myriad forms of life for billions of years before the arrival of humans is humbling. Is one to believe, as thoroughgoing anthropocentrism requires, that these beautiful and complex creatures had no value because there were no humans to value them?[39] Given the dimensions of the social and ecological crises humanity faces in the twenty-first century, the total rejection of this worn-out, arrogant anthropocentrism is sorely needed. As the eminent naturalist Aldo Leopold (1886–1948) once said, the extension of moral consideration to the "biotic community" is "an evolutionary possibility and an ecological necessity."[40]

The evolutionary history of *Homo sapiens* fundamentally challenges our anthropocentric paradigm and should be truly humbling. Yet, in many ways, the full ethical significance of Darwin's "dangerous idea" has not penetrated the popular consciousness. To a significant degree, our ethical comportment has not caught up with our metaphysical outlook. That is, although many people intellectually assent to the truth of evolutionary theory, our moral and legal systems reflect the assumption that humans alone have intrinsic value.

The U.S. legal system embodies the same simplistic, binary thinking advanced by Descartes. Before the law, an individual is either (1) a person, with attendant rights and responsibilities, or (2) property, and may be treated as such. This was seen most clearly in the infamous 1857 Supreme Court case *Dred Scott* v. *Sandford*, in which the court concluded (by a 7–2 margin) that Scott was not a

---

39. Biologist Stephen J. Gould had a memorable, if hyperbolic, way of responding to such a view. "Nature does not exist for us, had no idea we were coming, and doesn't give a damn about us." "The Golden Rule: A Proper Scale for Our Environmental Crisis," *Natural History* 99, no. 9 (September 1990): 24.

40. Aldo Leopold, "The Land Ethic," in *A Sand County Almanac* (New York: Ballantine Books, 1970 [1949]), 238–39.

person; he was property. As property, Scott was no more justified in refusing an order than would be a tractor. Although the *Scott* decision was superseded by the passage of the Fourteenth Amendment to the U.S. Constitution, the legal system continues to hold *Scott*'s binary logic at its core: either one is a person or one is property. Thus, while there are many laws restricting what one can do with one's property (e.g., animal welfare laws), these laws do not grant the property itself any rights. Paradoxically, then, although one can be fined or even imprisoned for torturing a dog (e.g., in 2007 the athlete Michael Vick was jailed for twenty-three months for abusing his dogs), the dog itself remains property and does not have a legal right not to be tortured. Why is it that our actions and laws maintain a dualism that is denied by our scientific theories?

As biologist Lynn Margulis, winner of the National Medal of Science, puts it,

> Even cosmopolitan thinkers, evidence-seeking scientists, and scholars who criticize excessive tribal loyalty do not necessarily repudiate, or even recognize, their anthropocentrism. Most educated people admit there is overwhelming evidence that we belong to the mammalian order of primates, but they still believe that we humans are the "highest" species of animals. Even as the Bible maintains the Jews to be the "chosen people," humans take it to be self-evident that "people are superior to all other life-forms."[41]

This belief conflicts with the world revealed by Darwin and contemporary evolutionary biology. Evolutionary theory teaches that the human animal is not a grand *exception to* the physical and biological forces that have shaped the natural world, but rather a beautiful *exemplification of* these forces.[42]

---

41. Lynn Margulis, "Gaia and Machines," in *Back to Darwin: A Richer Account of Evolution*, ed. John B. Cobb, Jr. (Cambridge: Eerdmans, 2008), 170.

42. As Margulis rather dramatically puts it, "Those twin delusions of human grandeur, our natural superiority and unique scientific objectivity, are spectacularly successful strategies for human survival. But they are illusions, even if scientists and scholars share them" (ibid., 172).

DISCUSSION QUESTIONS

1. How does evolutionary theory challenge older views that posit a hierarchy within nature?
2. What is the difference between progress and evolution?
3. What is the significance of the fact that we aren't zombies? How does this challenge mechanistic accounts of biological evolution?
4. What does it mean to say that the difference between humans and nonhumans is ultimately one of degree, not kind? What does it mean to say that humans are a grand exemplification of evolution, not a singular exception to it?
5. Does the fact that we evolved from simpler organisms affect our duties to nonhumans?
6. Given the vast history of the universe, does it make sense to say that humans are the only beings with intrinsic value? Did the myriad forms of life inhabiting Earth for billions of years before the evolution of *Homo sapiens* have any intrinsic value? Consider these questions in relationship to the statement in the book of Genesis that God created each thing and "saw that it was good."

GLOSSARY

anthropocentrism. Literally means "human-centered." It is the view that all meaning and value is derived from an entity's relationship to human beings; an entity has no value outside its relationship to human beings.

atom. For the ancient Greeks, the atom is the indivisible building block from which everything is made. From the Greek word *atomos* or *a-tomos,* which literally means "not cuttable" or "indivisible."

determinist (determinism). A determinist posits that every event in nature is determined by absolute laws of nature; thus, according to this view, there is no genuine novelty, much less freedom.

dualist (dualism). A dualist posits that reality is ultimately divided into two fundamentally different parts. Plato, for instance, was a dualist because he believed that the world was divided into this temporary, physical world and an eternal, unchanging world of ideas or perfect forms. Descartes was a dualist in that he divided reality into mind or "thinking things" and material or "extended things."

materialist (materialism). A materialist posits that reality is composed solely of matter. Materialism thus denies the possibility of all nonmaterial forms of reality (e.g., a nonmaterial mind or soul, a transcendent God).

***telos* (teleology, teleological).** An ancient Greek term referring to an end, aim, or purpose.

***psychē* (pronounced soo-khay).** Greek term referring to an internal principle of organization and change within any living thing that arranges its matter, defines what it is, and is the basis of its powers and capacities. Plants, animals, and humans (as living things) all have different kinds of *psychē*. Often translated (misleadingly) as "soul."

vivisection. Dissecting an animal while it is still alive.

RESOURCES FOR FURTHER STUDY

Ayala, Francisco J. *Darwin and Intelligent Design.* Minneapolis, MN: Fortress Press, 2006.

Ayala, Francisco J. *Darwin's Gift to Science and Religion.* Washington, DC: Joseph Henry Press, 2007.

Cobb, John, Jr., ed. *Back to Darwin: A Richer Account of Evolution.* Cambridge, UK: Eerdmans, 2008.

Dawkins, Richard. *The Blind Watchmaker.* New York: W. W. Norton, 1986.

Dawkins, Richard. *The Greatest Show on Earth: The Evidence for Evolution.* New York: Free Press, 2009.

Dawkins, Richard. *River Out of Eden.* New York: Basic Books, 1995.

Dennett, Daniel C. *Darwin's Dangerous Idea: Evolution and the Meaning of Life* (New York: Simon and Schuster, 1995).

Haught, John F. *Deeper than Darwin: The Prospect for Religion in the Age of Evolution.* Boulder, CO: Westview Press, 2003.

Haught, John F. *God After Darwin: A Theology of Evolution.* Boulder, CO: Westview Press, 2008a.

Haught, John F. *God and the New Atheism: A Critical Response to Dawkins, Harris, and Hitchens.* Louisville, KY: Westview Press, 2008b.

Haught, John F. *Making Sense of Evolution: Darwin, God, and the Drama of Life.* Louisville, KY: Westview Press, 2010.

Huchingson, James, ed. *Religion and the Natural Sciences.* New York: Harcourt Brace Jovanovich, 1993.

National Academy of Sciences. *Science, Evolution, and Creationism.* Washington, DC: National Academies Press, 2008.

Rolston, Holmes III. *Science and Religion: A Critical Survey.* New York: Random House, 1987.

Ruse, Michael. *Can a Darwinian Be a Christian?* Cambridge: Cambridge University Press, 2000.

Ruse, Michael, and Joseph Travis. *Evolution: The First Four Billion Years.* Cambridge, MA: Harvard University Press, 2009.

Turner, J. Scott. *The Tinkerer's Accomplice: How Design Emerges from Life Itself.* Cambridge, MA: Harvard University Press, 2007.

Whitehead, Alfred North. *Science and the Modern World.* New York: Free Press, 1925.

chapter 4

Theology in the Context of Evolution

By Rodica M. M. Stoicoiu
Mount St. Mary's University, Emmitsburg, MD

The preceding chapters have examined the biblical accounts of creation, the theory of evolution, the scientific method, and the history and philosophical foundations of the theory of evolution. This chapter explores the relationship between evolutionary science and faith in a good and loving God.

Theology means the study of God (from the Greek *theo* for "God" and *logy*, "to study"). Anselm of Canterbury (1033–1109), the renowned eleventh-century Christian philosopher and theologian, regarded theology as faith searching for understanding. Many other Christians consider theology the disciplined reflection on living the Christian life, as guided by the Bible, the liturgy, and church councils and creeds. Theology always occurs in a context of living persons, their history, and their culture. So, even as the core elements of Christianity remain steadfast (e.g., the Bible), for any of these elements to have life, they must be understood in the context of the person of faith. In other words, theology engages people as they seek to understand God in their individual lives and in their world. While theology has many branches, all theology asks questions about God and tries to build a deeper understanding of God in relation to people. Hence, questions regarding the relationship between God and science, and by extension God and evolution, are very much within the theologian's terrain.

Not only *can* theology ask questions of science, it *should* do so. Such questions engage the intellect as well as the heart; they are matters of both reason and faith. Contrary to what many well-intentioned religious and scientific leaders claim or imply and what the popular media emphasize, there is no contradiction between reason and faith. Belief in God does not require one to close one's mind. It is possible to have a reasoned faith in God while engaging the intellect. While faith is not the same as knowledge, neither is it irrational, magic, or a substitute for knowledge. Indeed, one must have an "informed faith," not just "take it on faith": faith should not be "blind."

Faith, then, draws on both the mind and heart (or cognitive and affective dimensions) in a personal response to the revelation of God. One would not enroll in a college one knew nothing about, so why would one believe in a God one knew nothing about? Faith and reason are partners. And because faith and reason work together, it is not only acceptable but necessary to ask questions of the world around one in order to know oneself and God more deeply. Thus, it is not a question of either believing in God or accepting the theory of evolution. One may believe in God *and* accept the theory of evolution, for both ideas address the reality of the universe, though in different ways. If the theory of evolution is true, then faith will agree with its conclusions.

From the moment Darwin's *Origin of Species* was published, people began asking questions about the relationship between evolution and theology. Some held the theory of evolution as proof positive that God does not exist, while others rejected evolutionary theory outright, arguing that God alone is responsible for the complex world and that evolution has no role in it. This chapter considers both responses to evolutionary theory and then moves beyond them in an attempt to offer the reader rational justification for holding both the theory of evolution and faith in God together.

Faith and science are not contradictory. They are different, however: Science is a matter of reason, of using the human intellect to hypothesize, experiment, and theorize based on evidence and inductive reasoning. Questions of God are questions of faith; they do not lend themselves to scientific inquiry, but this does not mean they are unreasonable. Faith by its very definition implies trust in another. Faith is both intellectual and experiential. That is, faith leads a person to live out that trust in a specific way. For Christians this means trust in God through Christ and living out that trust in a very real, experiential way among a community of believers that follow certain tenets of belief. Theology enters into the picture when, as noted, faith seeks understanding. This is accomplished by engaging the intellect so that one may come to a deeper understanding of the experience of God. Hence faith is rational, requiring intellectual assent. That faith and reason be held together is not only possible, it is preferable. It is far better to engage one's mind and to make use of one's intellect, so that reason and faith can engage and enrich each other. This approach pertains no less to questions of evolution.

EVOLUTION AND RESPONSES TO EVOLUTION

As noted elsewhere in this book, evolution can be understood simply as change over time caused by the process of natural selection.[1] Evolutionary theory holds that through a process of natural selection, those individuals most capable of living under certain conditions will reproduce more often, passing on more of their traits to the next generation (see the example of long-legged deer in chapter 2). Natural selection holds that the genetic variation among members of any given species will equip some members to better adapt to their environment, increasing the likelihood they will survive and pass on their genes to their offspring. The theological question that arises from this reality is, where in this process of selection is God? Truly understanding and accepting that the world around us is the result of evolution may challenge preconceptions of God and the role God plays in the world. If evolutionary change usually occurs slowly over long periods of time by way of natural selection, can one really speak of a divine plan? How can one speak of a good and loving God in a world formed by a process of selection that seems to lack any compassion? How is one to answer for the seeming brutality of existence in the face of evolutionary theory? If, as seems the case, the theory of evolution explains the diversity of life and human existence, then

1. The modern synthesis of this theory includes genetics and its role in evolution, something not explored in Darwin's time.

what need is there for God? What place does humanity have in such a world? And what about the biblical accounts of creation? How is one to understand these? Before discussing how God can be understood in relation to evolution, one must first address views that either deny evolution and defend God or defend evolution and deny God; such views include **creationism**, **intelligent design**, and **scientific materialism**.

CREATIONISM

Let's begin with creationism. First, creation must be distinguished from creation*ism*. The doctrine of creation recognizes the first three chapters of Genesis as a religious, not a scientific, text that focuses on the relationship of God with the world. As a religious text, this first book of the Bible says something about one's relationship with God; its intent is clearly religious, not scientific, a point explored thoroughly in chapter 1. The authors of the Bible are not describing the historical, scientific context for the existence of the world. Rather, the intent of Genesis is to show God as the Source of all there is and to show that this creation is good. "And God saw that it was good" is the constant refrain in the first chapter of Genesis, even as Genesis notes that human beings are made in the image of God and are themselves good and endowed with free will. In the third chapter of Genesis, human beings choose freely to eat of the fruit of the tree of knowledge, indicating that evil in the world is the result of human choice in the face of this free will; hence, evil is the responsibility of human beings.

Creation*ism*, on the other hand, holds that Genesis is a precise account of God's creation of the world. Creationists view Genesis as absolutely inerrant, scientifically and historically, as well as religiously. This is a position usually held by Christian fundamentalists. Such a reading rejects Darwin's theory of evolution as not just false, but dangerous. This view of evolution and Genesis sits squarely at the center of some of the most contentious battles in today's religious culture wars. These battles occur when religious groups fight for the inclusion of religious material in nonreligious venues, such as science classes. Two examples of this are the legal battles over whether to teach Genesis in high school science classes and similar attempts to introduce intelligent design in school curricula.

SCOPES MONKEY TRIAL

IMAGE: © BETTMANN/CORBIS

Clarence Seward Darrow (1857–1938), prominent American lawyer (left), sits next to William Jennings Bryan during the Scopes Monkey Trial of 1925. Darrow defended John Scopes, who was charged with violating a state law forbidding the teaching of evolution in public schools. Bryan served as prosecutor in the case, won by Bryan.

The Scopes Monkey Trial of 1925 presumed to put evolution on trial. John Scopes, a small-town biology teacher in Tennessee, was sued for teaching the theory of evolution, a crime in the state at the time. Defending him was Clarence Darrow, one of the top defense lawyers of his day. The trial made national news, and though Scopes lost the case in the court of law, he won in the court of popular opinion. Yet this battle continues today. A case in point involves the Dover, Pennsylvania, public school board, which in 2005 required the teaching of intelligent design in biology classes as an alternative to evolution. Parents, backed by the ACLU, filed suit against the school board. This lawsuit is known as *Kitzmiller, et al v. Dover School District, et al.*

According to creationism all living things are directly created by God in the order described in the first three chapters of Genesis. Thus, there can be no new species unless they come from God, and certainly if there are new species, they did not "evolve" from older ones as the theory of evolution would have it. Rather, they come into existence full blown from the hand of God.

Such thinking applies not only to human origins but also to the question of salvation. For creationists, evil in the world is the direct result of the original sin committed by Adam and Eve. The evil introduced into the world by this original sin required the intercession of a savior, without whom humanity could not be redeemed. Hence, if human beings arose through a process of evolution, there would be no original sin and no need for a redeeming savior; creationists therefore reject evolutionary theory.

One branch of creationism is *scientific creationism*, which contends that Genesis is "scientifically authoritative."[2] More often than not, those who hold this position are trained scientists who claim to apply the scientific method to biblical texts. As believers in **strict inerrancy**, they argue the Bible is literally true and cannot contradict science.[3] They take the Bible to be scientifically correct and reject any and all Darwinian claims of evolution.[4]

Scientific creationists claim that evolution is "only" a theory and hence cannot be taken seriously. That any scientific theory requires tremendous research and testing to earn the honorific *theory* (see chapter 2) is disregarded. They accept the creationist argument that the biblical texts themselves are accurate sources of scientific data. Because scientific creationists deem their own claims as "scientific," they feel justified in demanding equal time with evolution in the science curricula of public schools, a phenomenon particularly evident in the United States where the movement originated. Scientific creationists contend that students should be able to choose between alternate scientific theories. Yet, not only is the creationist position wrong in

---

2. John F. Haught, *Responses to 101 Questions on God and Evolution* (New York: Paulist Press, 2001), 72.

3. Ibid.

4. The opposite of this position is called ***limited inerrancy***, a position that accepts there can be scientific and historical error in scriptural texts. The truth in the text is what it conveys regarding salvation. This is discussed in chapter 1, page 3.

terms of its treatment of biblical texts, but as John Haught notes, "It also trivializes the sacred texts by bringing them down into the same secular context as modern scientific discourse. It forces biblical texts into a competitive encounter with scientific treatises, and in doing so suppresses any religious meaning they might have."[5] Theologians and scientists alike object to the teaching of scientific creationism in science classes, arguing that it is a violation of the theological integrity of the Bible.[6] (Recall the discussion of these texts in chapter 1.) The purpose of biblical texts is to make religious claims about God and humans' relationship with God. The writers of Genesis knew nothing about modern scientific method and were not making scientific claims about the world or about God. To deny this violates the intent of the authors of the Bible and the content of their texts.

INTELLIGENT DESIGN

Another approach taken in response to evolutionary theory is that of intelligent design (ID). Proponents of this argument believe that the intricacy of the universe requires the hand of an intelligent designer, God. When they look at the world, they see patterns and life-forms that they claim are too complex to have arisen by natural selection alone. In the face of the seemingly random nature of evolution, proponents of ID opt for an ordered, planned cosmos and an ordering, planning divine creator. Intelligent design in its current form is attributed to Phillip Johnson, a law professor who challenged the theory of evolution in a series of books in the 1990s. Johnson denies that what he calls "unintelligent, purposeless, natural processes" could be solely responsible for the universe.[7] He sees evolution as a system that seeks to deny a need for God in the process of creation, thus Johnson's battle to prove an intelligent designer lies at the heart of development of the universe.

Another important figure in this movement is Michael Behe, a Catholic biochemist who, in his book *Darwin's Black Box: The*

5. Haught, *101 Questions*, 73.

6. Ibid.

7. NOVA, *Defending Intelligent Design: An Interview with Phillip E. Johnson*, Public Broadcasting System, 2007.

Biochemical Challenge to Evolution, testifies to the molecular complexity of certain biological systems and concludes that God must be their source and designer.

One branch of intelligent design argues for **irreducible complexity**[8] (see also chapter 2). This variation on ID contends there are numerous examples in nature(e.g., the structures of a flagellum or the process of blood clotting)[9] that defy an evolutionary explanation. This view argues that these natural structures or organisms are so highly complex and interdependent, they could not have developed in stages over time. Instead, such structures require a creation that is whole and functioning all at once. Such structures and organisms, the argument for irreducible complexity concludes, must be the work of a divine planner or designer. This is essentially an argument for order, grounded on the belief that the universe is inherently organized and the only sensible explanation for its order is intelligent design.

Investigating more closely, however, one finds that neither a flagellum nor the process of blood clotting (to continue with these examples) is irreducibly complex.[10] Both can be explained by evolutionary theory. "Random variations, the mechanism of natural selection, and an ample amount of time (millions and millions of years) to filter out the adaptive from the nonadaptive features in organisms" sufficiently explains the development of intricate biological structures.[11]

Intelligent design is neither good science nor good theology. The scientific claims it offers are better explained by Darwinian evolution. In addition, the tendency of intelligent design to insert theology into scientific research diminishes one's view of God and the universe, by reducing God to a rigidly controlling designer and the universe to a well-organized, precisely ordered reality. ID also minimizes the role of science, not allowing science to function as intended, in a "naturalistic (methodologically agnostic) way."[12] There is no place in science

---

8. John F. Haught, *God After Darwin* (Boulder, CO: Westview Press, 2000), 3.

9. Ibid., 4.

10. Francisco Ayala, *Darwin and Intelligent Design* (Minneapolis: Fortress Press, 2006), 82.

11. Haught, *101 Questions*, 85.

12. Ibid., 89.

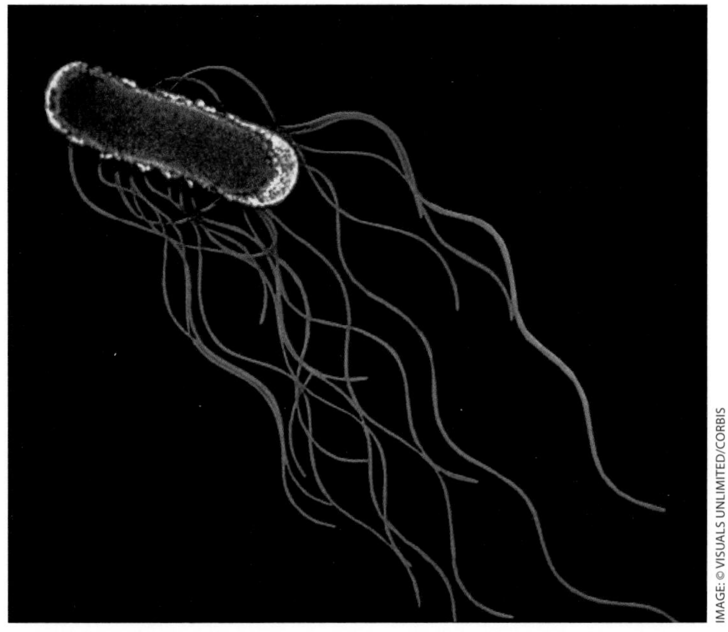

IMAGE: © VISUALS UNLIMITED/CORBIS

Salmonella bacteria showing its peritrichous flagella used in locomotion. Magnification of x 13,250.

for divine claims because science is methodologically naturalistic. Intelligent design interferes with the conduct of good science and lowers God by making the Divine merely one more cog in the wheel of scientific explanation, rather than the "depth and ground of all natural causation."[13] In other words, rather than recognizing God as the Source of all, God becomes just another explanation in a long list of explanations.

SCIENTIFIC MATERIALISM

There are scientists who see in evolution a denial of God's existence. For such scientists, known as scientific materialists, evolution is an "inherently meaningless process" that involves "random genetic

---

13. Ibid.

mutation, the deterministic laws of 'natural selection,' and enormous spans of time."[14] There is no purpose and hence no God. According to this view, evolution occurs in the midst of constant genetic rearrangement that provides some living organisms with a greater ability to survive and reproduce than others.[15] Over the enormous span of time through which these processes have occurred, evolved the variety of life on Earth today. No divine plan or divine planner is needed. Rather, as Richard Dawkins states, "blind chance and natural selection working over long periods of time can account for life's creativity all by themselves."[16] For some who make this argument, evolution is about the "selfish gene." Individual organisms are the "vehicles" in which these genes rearrange and reproduce themselves, and pass on their information to the next generation.[17] Dawkins argues that "genes will seek to survive at all costs and by whatever means available," including producing human beings, if this means a greater chance of survival.[18] In his view, there is no need for God.

For scientific materialists, the mindless mechanism of evolution is sufficient to explain the diversity of life on Earth.[19] Even some scientists who consider themselves faithful Christians take this stance, as seems to be the case with noted biologist Francisco Ayala, who writes, "it was Darwin's greatest accomplishment to show that the directive organization of living beings can be explained as the result of a natural process, natural selection, without any need to resort to a creator or other external agent."[20] For Ayala and other biologists who embrace such a viewpoint (this position seems to put him in the "separatist" camp—a stance discussed later in this chapter), Darwin's theory of evolution changes the place that human beings hold in the world. Their arguments view science alone as sufficient to explain the universe. There is no need to speak of God.

---

14. Haught, *God After Darwin*, 11. Here Haught is referencing the works of American philosopher Daniel Dennett and British zoologist Richard Dawkins, both of whom view evolution as proof that God does not exist.

15. Ibid., 12.

16. Ibid., 13.

17. Ibid.

18. Ibid., 14.

19. Ibid., 17.

20. Ibid., 18.

Scientific materialists work from a specific metaphysical viewpoint. Everyone holds a metaphysic, or a collection of conclusions or operative assumptions about the nature of reality. The metaphysic of scientific materialism claims that matter is "all there is," and there is no need of anything more. Rather, the universe is without purpose or direction.[21]

A note of caution is merited here. One must not leap to the conclusion that the Darwinian theory of evolution is the same as scientific materialism. It is not. Scientific materialism is a philosophical commitment that underlies a particular interpretation of the theory of evolution. Scientific materialism is an assumption rather than a scientifically provable fact. As we will discuss below, there are other ways of viewing the theory of evolution that are far more open to theistic interpretation than scientific materialism.

Scientific materialism sees the process of evolution as the purposeless reorganization of genetic material: From time immemorial this aimless reshuffling of genes has produced some organisms better fitted to survive than others, creating a progression of organisms that could continue for eons. All that one can be certain of is anchored in the past, which brought forth the current state of being, "step by fateful step." There is no room in this metaphysic for novelty, for unprecedented possibility. Indeed, according to scientific materialism, "the entire life-process, rather than being evidence of nature's openness to the arrival of genuine novelty, is only the explication of what was fully latent already in lifeless matter from the time of cosmic beginnings."[22]

GOD OF THE GAPS

Accepting that it is possible to agree with the scientific theory of evolution and still have faith in a God who is the ground of all being, begs the question of how one is to speak of the relationship between evolution and God. One common pitfall to avoid is to explain the role of God in relation to evolution as a "God of the gaps." In this

21. Ibid., 26.

22. Ibid., 86. The reader may also wish to refer back to the discussion of a mechanistic view of reality in chapter 3.

approach, which seeks to find a place for God while accepting the truth of evolution, God is used as a default explanation for those times when science fails to provide an answer. Into such gaps God is poured.

There are a number of problems with this view. For one, it relegates God to only one among many natural causes. This reduces God from the mysterious "ground of all natural causation" to merely a stopgap used to explain events or phenomena that science has yet to understand. In other words, when we are scientifically stumped, God is the default answer. The God-of-the-gaps explanation is a "science-stopper," inhibiting scientists from seeking a more adequate explanation. Why look for another explanation when we can throw God in at the end? One such "science-stopper" is the argument for irreducible complexity discussed earlier. The God-of-the-gaps response simply substitutes a supernatural answer for a scientific one. "It makes divine action one link in the world's chain of finite causes rather than the ultimate ground of all natural causes."[23]

SEPARATISM

Another view to be avoided that attempts to correlate evolution and God is known as **separatism**. Separatists belong in the same camp as those opposed to the theory of evolution. This is a fundamentalist position in which the truth of evolution and faith in God are both safeguarded by keeping the two strictly separate. "Separatism claims that science as such is self-consciously limited to dealing with questions about the physical or mechanical causes of events, whereas theology, by definition, is more concerned with questions about the meaning and ultimate explanation of things."[24] In other words, science does not deal with the supernatural, but theology does; and because the two deal in two totally separate areas, there is no reason for their paths to cross. This seems to be the position of Ayala, who writes that "science and religion concern nonoverlapping realms of knowledge."[25]

---

23. Ibid., 19.

24. Haught, *God After Darwin*, 28.

25. Ayala, *Darwin and Intelligent Design*, 91.

There appears to be a certain naïveté in this approach. It assumes that no questions of theological interest can be raised by science. But is this true? As will be explored below, the question of suffering—a reality made present through evolutionary processes—is of deep theological concern. According to separatism, though, theology cannot address this concern in terms of the role of evolution. Likewise, questions raised by the theory of evolution regarding the existence of God, questions raised by those arguing for the blind chance of natural selection, questions that are the source of much conflict today regarding theology and science, cannot be posed by separatists because of the line they draw between evolution and theology.

The theological separatist would respond to such questions with the following three arguments:

1. Natural selection is a purely mechanistic process that relies on blind chance, which implies that evolution is not meant to answer deeper questions of being, such as the reason for suffering.[26]
2. The suffering that results from evolutionary processes leads to positive growth and evolution is as good a school for such growth as any other.
3. The laws of evolution do not contradict the existence of God because questions about God are theological and hence belong to a different field altogether.

Such arguments maintain a distance between science and theology and fail to engage questions of God in the process of evolution itself.[27] Scientific and theological separatism divide scientific from theological issues and treat each as distinct fields, a rather safe road to take. A different approach is warranted, however, one that, as demonstrated below, asks difficult questions about God from within the theory of evolution itself.

Creationism denies evolution, scientific creationism attempts to apply scientific principles to biblical accounts to provide a parallel explanation, and scientific materialism refutes any theological claim

26. Haught, 29.

27. Ibid.

with its antitheistic metaphysic. Unlike these views, separatism steers clear of all ideological responses and hence takes no issue with either evolutionary theory or faith in God. However, by refusing to directly engage evolutionary theory and faith, it fails to achieve a depth and richness of response.[28]

SUFFERING

The question of suffering is keenly important in theology. The question is asked, if God is a good and loving Creator, why would God's creation include suffering and pain? The theory of evolution by virtue of its random nature makes room for suffering. Indeed, when one considers the millennia of suffering of countless species—most of which are now extinct—the issue of **theodicy** or the defense of God's goodness in the face of evil is brought into stark focus.[29] Evolution challenges theology to confront the question of suffering.

One response is to simply deny that God exists, for what God would allow "every leaf, blade of grass, and drop of water to be a battleground in which living organisms pursue, capture, kill and eat one another?"[30] Evolution reveals a world in which the survival of the fittest involves suffering, pain, and death, especially for those less fit. Yet this does not describe the totality of the world that evolution has fashioned. Even as there is suffering, cruelty, and pain, evolution brings companionship, cooperation, interdependency, beauty, and compassion.[31] The theory of evolution has made a difficult situation even more contentious. How is theology to respond to the challenge Darwin presents?

The Christian response to suffering is framed by the cross of Christ. Christianity teaches that Christ suffered on the cross for the sake of all human persons. So it is through Christ's suffering that Christians come to understand their own. In the midst of the terrible suffering human beings inflict upon each other, and in the face of the millions of life-forms that fossils show have lived and suffered on

---

28. Ibid., 31–32.

29. Haught, *101 Questions*, 123.

30. Haught, *God After Darwin*, 21.

31. Ibid., 45.

Earth over countless eons, Christians find hope in Christ's suffering. As theologian David Power states,

> As Christ entered the tomb to be among those who dwelt in the absence of God, so it is in passing through hell, the place of the dead, that we hear God speak and we hear the word of Jesus who speaks to us of the one of whom he is Son—of the God who is Father, not Judge or Almighty or Law—and who has given the divine Spirit without calculation of any sort to those whom he chooses to call children.[32]

Through the suffering of Jesus, God incarnate, one may recognize "a God who participates fully in the world's struggle and pain."[33] This poses a challenge to many who imagine God as "Judge" or "Almighty," rather than as a loving and caring parent. Evolution opens the door for one to appreciate the compassion of a God whose caring extends over countless eons, even as it reveals new ways of understanding the claims that Christian tradition and texts have always laid out.[34]

The image of a caring God who suffers alongside creation—rather than a distant, even cold, creator—conflicts with some popular conceptions of God. Such a compassionate and loving image has a long history in Christianity, however. The earliest Christians understood God in this way. This is the Incarnation, God become human, unto death on a cross. This understanding of a gentle, close God is conveyed in the images of Christ as a shepherd boy painted on the walls of early Christian tombs. This is the ministry of Jesus, who cared for the sick, the poor, and the disenfranchised of his time. It is in Christ's suffering, death, and Resurrection that all Christians find hope. This image of God only changes when Christianity becomes

---

32. David N. Power, *Love without Calculation: A Reflection on Divine Kenosis* (New York: Crossroad Publishing, 2005), 2.

33. Haught, *God After Darwin*, 46.

34. Ibid. Haught uses **process theology** to recognize a God who walks with rather than dominates creation. Process theology posits a God who is intimate with creation, not distant and cold. Instead, the universe is in process, and this process, with all of its experiences, contradictions, sufferings, etc., is "harvested into the divine experience. . . . Here all the suffering, struggle, loss and triumph in evolution are finally endowed with eternal meaning" (Ibid., 128). The result is a universe in which suffering is folded into God's ongoing, compassionate embrace and, hence, a universe that is not without meaning or value.

IMAGE: © ELIO CIOL/CORBIS

Jesus Christ as the Good Shepherd. From an early Christian floor mosaic at the Basilica of Aquileia.

the religion of empire. Then Christ is no longer depicted in humility but in power, as lord over all, head of the heavenly hierarchy, divine emperor.[35] The theory of evolution can make one more receptive to a humble God, a God who is close to people, who can understand human suffering because he shared it on the cross.

This image of a God who empties self unto death on a cross and who shows his greatest power in becoming defenseless is helpful as one tries to explain suffering in the face of evolution. Another word for such self-emptying is **kenosis**. There are three aspects of kenosis, a central Christian concept. The first aspect is understood as

> the presence [of Jesus] in the finitude of human existence in all its limitations and travail, but also in its joys,

35. Ibid., 48.

> companionship and hope. . . . The second aspect of *kenosis* is in [Jesus'] witness before the powers of the world, especially in his trial, passion and death. There he was reduced to humiliating forms. . . . The third aspect is death itself, knowing death to its fullness in order to show God's revelation of love and mercy in this death. It is in dying, not in avoiding death, to show himself superior to Hades, that Jesus, the Son and the Christ of God, overcame death. This is death among others, with others, for others. It is passing through death in order to enter life and to open the way for others to make the same passage.[36]

Christians may begin to make sense of the "blind chance" of evolution over eons of natural selection with its apparent "waste and suffering" by understanding God in this kenotic light.[37] It is an understanding of God manifesting self in weakness and vulnerability.[38]

The picture of an incarnate God who suffers along with creation is offensive to our customary sense of what should pass muster as ultimate reality. But perhaps this image can be called "revelatory" precisely because it breaks through the veil of our pedestrian projections of the absolute, and does so in such a way as to bring new meaning to all of life's suffering, struggle, and loss. This new meaning consists, in part at least, of the intuition that the agony of living beings is not undergone in isolation from the divine eternity but is taken up everlastingly and redemptively into the very "lifestory" of God.[39]

Christian faith has declared time and again that God showed God's love of creation by becoming a creature within it. Through the Incarnation, God through the Son embraced all of creation in an act of self-emptying love. And in embracing that creation, in Christ becoming a creature, God also embraced the suffering and pain of that creation along with the joy. This is the critical insight that Christianity brings to bear on the question of suffering raised by the theory of evolution.

---

36. Power, *Love without Calculation*, 47.

37. Haught, *God After Darwin* 49.

38. Ibid., 50.

39. Ibid.

EVOLUTIONARY THEOLOGY

So the question posed at the start of this chapter is once again raised: what is the relationship between the science of evolution and the theology of a good and loving God? Or, how can one engage theologically with the theory of evolution? This chapter has attempted to do just this by addressing many specific questions. Scientific materialism, which finds in evolution the perfect atheistic explanation of creation, has been examined. For adherents to this view, evolution leaves no room for God. Yet many Christians argue that theology has a rightful place in conversation with evolution, because theology (and hence God) provides answers to the ultimate "why" questions that evolution cannot answer. This chapter argues that evolution is not random and mindless, as scientific materialists argue, nor rigidly ordered by an intelligent designer, as ID proponents assert, but rather evolution is the result of God's "self-emptying suffering love."[40] But it is now time to take into account a still-fuller theological response to evolution. This response is posed in the theology of the Jesuit theologians Karl Rahner and Pierre Teilhard de Chardin and is presented here as interpreted by John Haught.

KARL RAHNER AND PIERRE TEILHARD DE CHARDIN

Karl Rahner, SJ (1905–1984)

Dr. Karl Rahner

Karl Rahner was one of the twentieth century's most prominent Catholic thinkers. A Jesuit scholar, his theology is described as "incarnational"; that is, he sought to understand God with and through an anthropology that was deeply respectful of the human person. Rahner was present at the Second Vatican Council as a theological expert and participated in the preparatory phase of the council and in the council itself. He

Continued

40. Ibid., 53.

RAHNER AND TEILHARD DE CHARDIN ***Continued***

taught at the universities of Munster and Munich and published more than 3,500 works in his lifetime. (Leo J. O'Donovan, ed., *A World of Grace: An Introduction to the Themes and Works of Karl Rahner's Theology* [New York: Seabury 1980.])

Pierre Teilhard de Chardin (1881–1955)

IMAGE: © BETTMANN/CORBIS

Pierre Teilhard de Chardin, SJ

Teilhard was a Jesuit with a deep interest in the natural world, having studied geology and paleontology and completed his dissertation on the geology of the Eocene epoch. His scientific interests and knowledge of evolution led him to consider ways to accept evolution while maintaining faith in God. Teilhard worked during a highly conservative time in the Church, and his writings were under close scrutiny for many years. He finally realized he would not be allowed to publish his work during his lifetime. Today his work is recognized as a vital contribution to theological thought, one that provides fresh insight into the relationship between theology and science, and in particular evolutionary science. (John and Mary Evelyn Grim, *Teilhard de Chardin: A Short Biography, http://www.teilhard dechardin.org/biography.html*)

Evolutionary theology can be summarized as follows. The world is in a process of becoming. This movement occurs throughout the cosmos by way of evolution—that seemingly chaotic method of natural selection that produces a display of life full of innovation and originality. In this process of becoming, Rahner argues, the world is drawing closer to the mystery of God by way of the "divinization of the world as a whole."[41] This could be termed an ***eschatological***

41. Karl Rahner, "Christology with an Evolutionary View of the World," in *Foundations of Christian Faith: An Introduction to the Idea of Christianity* (New York: Crossroad Publishing, 1989), 181.

theology, a theology focused on the future fulfillment of God's promise, when all that has begun reaches its completion in God. In other words, the world is in receipt of God's promise that all of the universe will reach its completion in the mystery of God.[42] God is both the ultimate Creator of the cosmos and also the Goal and Endpoint of the cosmos.

Teilhard took this focus on the future and made it the centerpiece of his own metaphysic and theology.[43] He rejects as incomplete and dangerously narrow the metaphysic of scientific materialism, which asserts that evolution negates the existence of God. Though some evolutionary biologists may disagree (mostly scientific materialists), Teilhard poses the metaphysical argument that the universe is evolving in the direction of "organized complexity."[44] He does not deny that evolution twists and turns in its particulars, but argues that the universe as a whole is "gathering the human mass towards a new future."[45] He posits that evolution not only brings forth abundant new realities over time but also moves toward increasing complexity. Haught notes that Teilhard's argument includes not just an increasingly complex biological reality but also an increasingly complex social reality in human society and technology.[46]

Even as Rahner posits that people are approaching a future fulfillment that will bring them into the mystery of God through a process of self-transcendence, Teilhard argues that God is the "overarching" force drawing all life toward completion: "Evolution . . . seems to require a divine source of being that resides not in a timeless present located somewhere 'up above' but in the future, essentially 'up ahead,' as the goal of a world still in the making."[47] This is quite a shift from the metaphysic familiar to most Christians. In the "classical" Christian metaphysic, the world is an inadequate reality that has fallen short of perfection; consider the Garden of Eden and the idea of heaven: all that is imperfect here is perfect

---

42. Ibid., 187.

43. Haught, *God After Darwin*, 82.

44. Haught, *101 Questions*, 134.

45. Ibid.

46. Ibid.

47. Haught, *God After Darwin*, 84.

there.[48] Such a metaphysic is as incompatible with evolution as scientific materialism is with theology.

This chapter has examined the relationship of evolution to God. It was noted that some scientists, utilizing a materialistic metaphysic, use evolution to argue against the existence of God, and that a closer look at materialistic metaphysics reveals ways scientists can defend the theory of evolution without resorting to atheism. The arguments of creationists and proponents of intelligent design that defend the existence of God by denying evolutionary theory have also been examined. Two arguments that hold God and evolution as compatible but do so in less than complete ways, that is, the God-of-the-gaps and separatist accounts, have been considered. The challenge of suffering raised by both evolution and theology has been explored. Finally, a theology that accounts for God and evolution, not as two subjects coexisting side by side, but as two mutually inclusive realities, has been presented.

In conclusion, the task undertaken in this chapter is theological, not scientific. And this theology argues that God is the ultimate ground of evolutionary science and the ultimate future toward which evolution is progressing. This is not a vision of God as judge or king, but of God as a gentle, close God, who persuades rather than commands, and whose creation is always in the process of becoming something new. In this sense, reality is a grand journey, and God is the ultimate adventurer.[49] In such an understanding, evolution is constantly open to new permutations, and while such a world tends not to be ordered but chaotic, it is also hopeful and directed to the future.

DISCUSSION QUESTIONS

1. Is it possible to maintain faith in God and accept the theory of evolution? Explain.
2. Explain why creationism and intelligent design might be insufficient responses to the theory of evolution. What arguments from this chapter might form a view that accepts both evolution and theology?

48. Ibid., 85.
49. Haught, *101 Questions*, 136.

3. Drawing on the discussion of evolutionary theology in this chapter, consider the concept of separatism in relation to science and theology.
4. People often ask how suffering is possible if there is a good and loving God. Discuss how evolutionary theology can address the issue of suffering, especially as it is encountered through evolutionary processes.
5. This chapter discusses three alternate views of creation without evolution. Outline two of these views. Explain any possible inadequacies in the views you have outlined. Try to formulate your own argument for creation without evolution.
6. Compare and contrast the arguments of Teilhard and Rahner presented in this chapter with one other argument presented in the chapter. What position would you take and why?

GLOSSARY

creationism. The stance that the account of creation in the book of Genesis is strictly inerrant. Adherents hold that the creation account in Genesis is scientifically, historically, and religiously accurate.

eschatological theology. Theology that focuses on the fulfillment of God's promise in the Kingdom.

intelligent design. The belief that the intricate complexity of the universe must be the work of a divine creator. Faced with what they perceive as the seemingly random nature of evolution, proponents of ID find patterns and order in creation, which they argue indicates an intelligent designer.

irreducible complexity. A variation on intelligent design that claims natural structures and organisms that appear too complex to arise by natural selection over time must have arisen whole all at once, thus requiring an intelligent designer.

kenosis. The self-emptying of God.

limited inerrancy. An approach to biblical texts that asserts they are inerrant in matters pertaining to faith but may contain historical and scientific errors.

process theology. Theologies that understand God as continually changing, hence the term *process*, God is in process. "God is the great companion . . . the fellow sufferer who understands."[50]

scientific materialism. The view that the process of evolution is mindless, random, and without purpose, and that this quality negates the existence of God.

separatism. A fundamentalist position that separates theological inquiry from science. Science is kept separate from faith.

strict inerrancy. The view that biblical texts offer scientifically and historically accurate accounts of creation; hence a strict inerrantist rejects Darwin's theory of evolution on the grounds that Genesis provides an accurate account of the creation of the world.

theodicy. The defense of God in the face of evil.

RESOURCES FOR FURTHER STUDY

Ayala, Francisco. *Darwin and Intelligent Design*. Minneapolis, MN: Fortress Press, 2006.

Dei Verbum (Dogmatic Constitution on Divine Revelation). Promulgated by His Holiness Pope Paul VI on November 18, 1965. *http://www.vatican.va/archive/hist_councils/ii_vatican_council/documents/vat-ii_const_19651118_dei-verbum_en.html*, especially chapter 3, no. 12.

Haught, John F. *God After Darwin: A Theology of Evolution*. Boulder, CO: Westview Press, 2000.

Haught, John F. *Responses to 101 Questions on God and Evolution*. New York: Paulist Press, 2001.

"The Interpretation of the Bible in the Church." *Origins*, January 6, 1994. *http://catholic-resources.org/ChurchDocs/PBC_Interp.htm*.

50. William Collinge, *The A to Z of Catholicism* (Lanham, MD: Scarecrow Press, 2001), 427.

READING REALITY

In these four chapters, each author has tried to present ways of "reading" reality that are faithful to their respective scholarly disciplines while insisting that these disciplines can, and must, stand in fruitful conversation with one another. If properly understood, it is possible to see that both science and religion can be successful ways of getting at the true nature of reality. In considering how one might fully and consistently affirm both science and religion, it may be helpful to ponder the following analogy.[1]

Imagine that a copy of J. K. Rowling's *Harry Potter and the Deathly Hallows* is lying open on the floor of a family room. The family dog, Porter, is curious and walks over to inspect. On his "reading," there is a white surface with many black marks. Finding nothing to keep his attention and after drooling a bit on page 217, Porter decides to curl up in his bed for a nap. A few moments later, 4-year-old Nora enters the room. Seeing the big book on the floor, she decides to inspect it too, noting not only the white surface and black markings but finding in them the familiar shapes of individual letters. Quickly bored with the big "chapter book," she decides to do some coloring in the other room. Next enters 7-year-old Hope, who, having heard about Harry Potter from older classmates, picks up the tome to see what the fuss is about. Not only does she recognize the familiar white pages with black letters, but she is also able to see that the letters combine to form words and sentences. Finding the vocabulary a bit too demanding, she sets the book back down, resolving to remember to ask her mother to read it to her. As she leaves the room, Hope passes Simon, a 14-year-old who quickly cradles his mother's book to pick up where he left off. Simon loves the suspense and adventure

1. This analogy is adapted from the work of John F. Haught, *Deeper than Darwin* (Cambridge, MA: Westview, 2003), 14–15. As Haught notes (192, n. 2), his version of the analogy is adapted from E. F. Schumacher (who, in turn, was following ideas from G. N. M. Tyrrell) in *A Guide for the Perplexed* (New York: Harper Colophon, 1978), 41–42.

as Harry, Hermione, and Ron struggle to fight He-Who-Must-Not-Be-Named, Lord Voldemort. After only a couple of chapters, Simon is called away to eat dinner. Later that evening, after all the kids are finally in bed asleep, their mother, Suzie, grabs her copy of *The Deathly Hallows* off the floor in hopes of finishing her second reading of the series. As she finishes the final chapters, she reflects that the book is not only a wonderful adventure story about the travails of Harry and his friends but also a penetrating commentary about the eternal struggle between good and evil, the nature of friendship, and the difficult search for genuine happiness.

Notice first that each of these readings is perfectly valid. Although Nora's reading of individual letters reveals a deeper meaning, her reading in no way falsifies Porter's "reading" of a white surface with black markings. Both are perfectly true. Similarly, Hope's reading of words and sentences is no less true than Simon's understanding of these sentences as revealing an adventure story or the mother's reading of the story as a commentary on the human condition. Though some readings may get at deeper levels of meaning than others, each level of reading gets at a true version of the "reality" that is the book.

Now imagine that trying to understand reality is like reading a book. Although it is certainly possible to "misread the text," in principle, the truth of one level does not negate the truth of the others. Physics, for instance, tells us about the most basic structures and regularities governing reality. It reveals, for instance, the way in which molecules of black ink become impregnated into the fibers of the white paper. Going beyond the properties of individual elements, chemistry explores the composition and interactions of different substances, just as Nora's reading of individual letters grasps a new level of meaning beyond Porter's seeing of white and black. Similarly, biology's study of integral living organisms reveals properties undiscoverable at the chemical or physical levels, just as Hope's reading and understanding of the meaning of words and sentences is not reducible to Nora's reading of individual letters. Each new level brings with it meaning and information that is not discoverable or reducible to the more basic levels. Thus, there are hundreds of different levels at which reality can be "read." Every discipline from physics to psychology and from sociology to theology is a "reading" that seeks to get at a certain level of meaning about reality. When understood at its proper level, each new layer deepens and enriches one's understanding and appreciation of the complex narrative that is the story of our universe.

This analogy is also instructive for considering the relationship between science and religion. Just as the mother finds that *Harry Potter and the Deathly Hallows* is more than an adventure story, and biology finds that the origin and structure of living organisms is more than just chemistry and physics, philosophy and religion find that the meaning and value of reality go beyond its material constituents. Science gets at *how* reality is structured while philosophy and religion get at *why* reality is at all; that is, they seek to understand the meaning and value of reality and human existence. The *how* and the *why* then become partners in humanity's ongoing search for truth in its myriad manifestations. Resisting the seductive tidiness of "simplistic readings" of reality offered by well-intentioned religious believers and science-minded secularists, it becomes possible to see that both science and religion reveal different parts of the "story" of our universe. To recognize this is not to arrive at a single, tidy conclusion, but to commit oneself to the search for a reasoned faith.

Brian G. Henning

ABOUT THE AUTHORS

Sr. Mary Katherine Birge, a member since 1982 of the Sisters of St. Joseph of Springfield, MA, teaches Scripture and theology at Mount St. Mary's University, Emmitsburg, MD. She received her MA in the classics from Tufts University; an MA in theology from St. John's University, Collegeville, MN; and her PhD, with distinction, in biblical studies from The Catholic University of America. Her scholarly interests range among the areas of Pauline literature, Synoptic Gospels, feminist hermeneutics, biblical spirituality, and Ignatian studies. She has spent most of her working life in the classroom, high school and undergraduate, and finds the daily give-and-take with students one of the most rewarding aspects of teaching.

Dr. Brian G. Henning earned his MA and PhD in philosophy from Fordham University. He is currently associate professor of philosophy at Gonzaga University, in Spokane, WA. Henning is author of the award-winning book *The Ethics of Creativity: Beauty, Morality, and Nature in a Processive Cosmos* and the coeditor of *Beyond Metaphysics? Explorations of A. N. Whitehead's Late Work*. His current scholarship and teaching focus is on the interconnections among ethics, metaphysics, and aesthetics.

Dr. Rodica M. M. Stoicoiu received her MA in theology from the University of Notre Dame and her PhD from The Catholic University of America. She is currently assistant professor of theology at Mount St. Mary's University, Emmitsburg, MD, where she teaches such courses as Foundations of Theology, Early Christian Thought, Sacraments, and Church Past and Present. Her writings include numerous articles on the theology of the Eucharist, the role of liturgy and social justice, and liturgical theology. Her current interests include the Catholic-Orthodox dialogue on liturgy and sacraments, and her current writing projects encompass the work of the dialogue and its effects upon the liturgical ecclesial nature of the Eucharist and Church Order.

Dr. Ryan Taylor received his MS in ecology from Florida International University and a PhD in evolutionary biology from the University of Louisiana at Lafayette. He is an assistant professor of biology at Salisbury University, Salisbury, MD. His research asks questions about animal communication systems and addresses the role that communication plays as a driver of evolutionary change. While he is broadly interested in evolution and animal communication, his current research uses the courtship of frogs as a model system to increase our understanding of evolution.

INDEX

Page numbers with special designators indicate illustrations (i), sidebars (s), figures (f), maps (m), captions (c), and footnotes (n).

P

R

S

T

V

W

Y

Z